MARINE PLANTS
OF THE CARIBBEAN

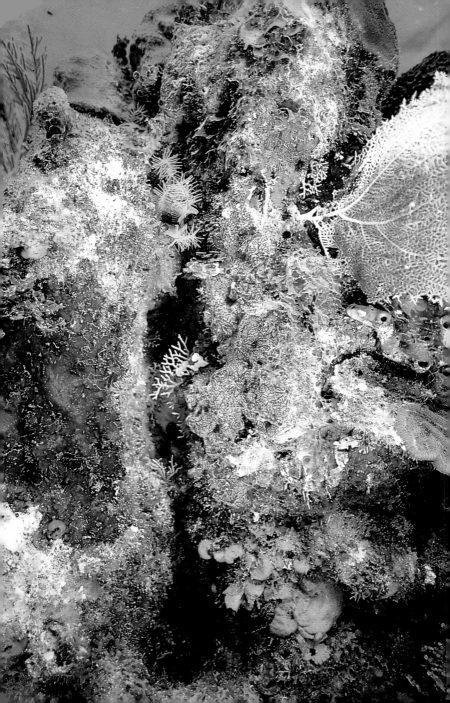

MARINE PLANTS
OF THE CARIBBEAN

A FIELD GUIDE FROM FLORIDA TO BRAZIL

Diane Scullion Littler, Mark M. Littler,
Katina E. Bucher, and James N. Norris

SMITHSONIAN INSTITUTION PRESS
WASHINGTON, D.C.,

Designer: Linda McKnight
Editor: Lorraine Atherton
Coordinating Editor: Leigh A. Alvarado

Library of Congress Cataloging-in-Publication Data

Marine plants of the Caribbean.
  Bibliography: p.
  Includes index.
  1. Marine algae—Caribbean Area—Identification.
2. Marine flora—Caribbean Area—Identification.
3. Marine algae—Caribbean Area—Pictorial works.
4. Marine flora—Caribbean Area—Pictorial works.
I. Littler, Diane Scullion.
QK572.2.A1M37 1989   589.39235   88-43157
ISBN 0-87474-607-8 (alk. paper)

British Library Cataloging-in-Publication Data
available

∞ The paper used in this publication meets the minimum requirements of the American National Standard for Performance of Paper for Printed Library
Materials Z39.48-1984

Color, text, and all other illustrations printed by
South China Printing Company, Hong Kong
Manufactured in Hong Kong

10 9 8 7 6 5 4 3 2 1
98 97 96 95 94 93 92 91 90 89

Table of Contents

Acknowledgments

We are indebted to Isabella Abbott, David Ballantine, German Bula Meyer, and Suzanne Fredericq, who critically examined all of our algal photographs. In addition, we thank Stephen Blair (*Halimeda*), Dennis Hanisak (*Codium*), John Kilar (*Sargassum*), John Reed (corals), Celia Smith (*Bostrichia*), and Robert Steneck (crustose corallines) for helping with specialized groups. Valuable technical assistance was provided by Barrett Brooks, Robert Sims, William Lee, Sherry Reed, and Kjell Sandved. Phillip Taylor and Paul Jensen provided helpful editorial comments. Thanks also go to William Fenical and Brian Lapointe for the support they provided on their National Science Foundation-sponsored cruises. Contribution No. 202 of the Smithsonian Marine Station at Link Port, Florida, and Contribution No. 217 of the Smithsonian Institution's Caribbean Coral Reef Ecosystem Project, Washington, D.C.

Introduction

Beyond the palms and sun-drenched beaches of the Caribbean, in the vast expanse of tropical blue water, lies the underwater world of the reef, a remarkably different place of wildly colorful plants and animals. Reefs are one of the few places where one can observe a complex community of plants and animals interacting naturally, seemingly little disturbed by human presence. Anyone who swims can enjoy this area by donning mask, fins, and snorkel; even beachcombers can find unique treasures along the white sand stretches adjacent to this fascinating undersea world.

This book has been designed for those individuals who wish to increase their knowledge of the extremely important but often overlooked photosynthetic organisms of the reef. This volume depicts 209 marine plants, including 204 kinds of algae and 5 seagrasses. A brief description of each species (the word "species" is both singular and plural), giving the size, shape, color, depth, and typical habitat is accompanied by an underwater photograph of the plant that emphasizes the prominently visible characteristics to facilitate identification. Scientific terminology has been kept to a minimum; when used, technical terms are either explained where they occur in the text or defined in the glossary at the back of the book.

We present the common species likely to be encountered, along with those that are locally abundant or might attract attention because of some unusual characteristic, and have not attempted to include all of the more than 600 species reported from the Caribbean Sea (Taylor, 1960; Wynne,

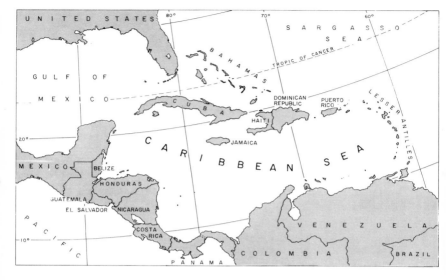

Figure 1. The Caribbean Sea and adjacent areas.

1986). Most of the plants selected can be identified without the aid of a microscope or other costly equipment. Many of the species included are distributed throughout the warm western Atlantic (Fig. 1), and thus this guide will be useful not only in the Caribbean Sea but also for a wide area including southern Florida, the Florida Keys, the Bahamas, and Bermuda.

MARINE PLANTS

Marine plants, unlike some marine animals, are limited in the depth to which they can grow by the penetration of light, the essential energy source for the process of photosynthesis. In the clearest of tropical waters, plants have been collected to depths of 268 meters (San Salvador Island, Bahamas); however, the largest diversity and abundance occurs from the intertidal zone to a depth of 30 meters. The depth ranges for the various species given in this book are those depths where the plants are most commonly found by the average snorkeler or diver. Consequently, certain plants may occur deeper or shallower than listed, depending on the surrounding environment.

Marine algae and seagrasses form the base of the oceanic food chain; they are the primary producers, converting sunlight energy and nutrients into plant materials, which provide food, oxygen, and habitats, directly or indirectly, for most other inhabitants of the seas. Nearly all marine plants serve as food for grazing animals such as snails, crustaceans, sea urchins, and fishes. Fish grazing, in particular, greatly affects the composition, distribution, and abundance of plant life on most tropical reefs.

Unlike seagrasses, which are flowering plants, marine algae lack conductive tissue and hence true roots, stems, leaves, and flowers. Consequently, we use terms such as "blade" or "frond" for the leaflike structure and "stipe" or "stalk" for the stemlike structure. Rootlike structures are referred to as rhizoids, whereas runners or stolons (as observed between strawberry plants) connecting upright blades are called rhizomes (Fig. 2). The word "algae" is plural, having a soft *g* and pronounced "al-jee"; "alga" is singular, with a hard *g* as in "gale."

Algae have many diverse ways to reproduce; sometimes several different mechanisms are present within a given species, nature's way of ensuring that one of the mechanisms will produce offspring. In general, though, algae reproduce by the release of spores (but lack the ability to produce a true seed) or by vegetative fragmentation (pieces breaking off to form a new plant). The types of reproductive structures that produce the spores can be important in making a proper identification. However, most of these structures are microscopic, so we will limit our discussion of this subject because of its complexity and because of the need for specialized stains and microscopes. Reviews of the reproductive methods and complex life histories of the algae may be found in

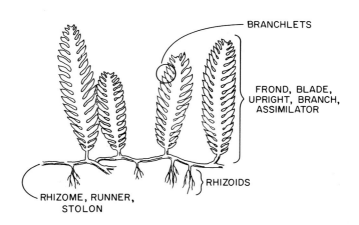

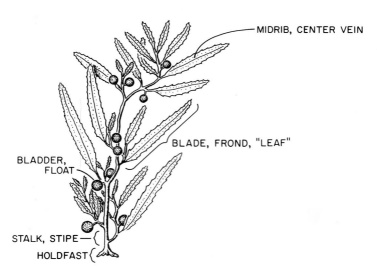

Figure 2. The major features of upright algae such as Caulerpa *(above) and* Sargassum *(below).*

texts by Bold and Wynne (1978), Lobban and Wynne (1981), and Sze (1986).

CLASSIFICATION AND FUNCTIONAL-FORM GROUPINGS

The descriptions in this volume are arranged by phylum. A phylum is a broad level in the scientific classification of organisms. Phyla (plural) are subdivided into smaller and smaller subgroupings; each phylum is divided into classes, a class into several orders (with names ending in "-ales"). An order is divided into families (ending in "-aceae"), a family into various genera, and finally a genus into species. The algae of the Caribbean are grouped into six phyla, with the name of each phylum being derived from the color of the dominant photosynthetic pigments. Of these, green algae (Chlorophyta), brown algae (Phaeophyta), and red algae (Rhodophyta) are the most conspicuous phyla present.

The three other algal phyla include forms of blue-green algae (Cyanophyta), diatoms (Chrysophyta), and dinoflagellates (Pyrrhophyta). The majority of diatoms are microscopic, so only one example has been included. A few species of marine diatoms form gelatinous colonies large enough to be conspicuous, but they usually break apart quite easily relative to other algae. The dinoflagellates are also microscopic; however, their pigments are responsible for the variety of color in many of the corals, where they live symbiotically (intimately associated) within the animals' tissues. Microscopic blue-green algae may occur in slimy gelatinous clusters, threads, or chains, often large enough to be conspicuous, but are not included in this guide because of their difficult and unsettled taxonomy and the necessity for microscopic examination. Only one important representative is described, because it is responsible for black-band disease commonly found on corals. The final plant group discussed includes the seagrasses, or marine flowering plants (Magnoliophytae, or Angiospermae), which contain vascular tissue and are easily distinguished from macroalgae by their stiffness, grasslike appearance, and grass green color.

The sizes and forms of algae are highly diverse, but certain basic features are common to many groups. It must be kept in mind, however, that although some specimens appear to be externally similar, they can belong to different taxonomic groups that are not closely related in their formal scientific classification. Conversely, closely related members of the same group (often based on microscopic reproductive characters) may bear little resemblance to each other in their external form.

The formal classification system is not only useful to specialists in the field but it is also helpful to someone interested in how the plant in hand is related to or different from others. Rather than following the conventional classification system, we have organized this book on a more obvious visual system, which we hope will be easier for interested naturalists. So that evolutionary relationships are not obscured for those with more technical interests, the family and order levels of classification are given with each description.

A more practical and visual means of comparing algae is that of functional-form groupings, which have proven to be closely linked with physiological and ecological characteristics and often indicate a great deal about a given plant's lifestyle. Consequently, throughout this book we have arranged forms together within each phylum; this facilitates comparison with similar forms and enables easier identification. The basic form groups range from delicate to coarse and are as follows: thin sheetlike and tubular forms, finely branched and delicate forms, coarse (usually branched) forms and saclike (balloon-shaped) forms, thick leathery and rubbery forms, articulated or jointed (hard but having flexible joints) forms, and crustose (a hard surface layer) forms.

The functional-form groups are arbitrary divisions in a continuum with one group overlapping the next; however, the extremes are unmistakably different from one another. The thin sheetlike and tubular forms are generally short-lived and fast growers. Their large surface area relative to volume enables the plants to have most of their cells exposed to sunlight for photosynthesis and in contact with the water for rapid uptake of nutrients. In general, thin forms put more energy into building photosynthetic material. Some are only one or two cells thick; consequently, they usually are fast-growing and the first plants to appear on newly exposed substrates (in other words, they are early sucessional plants and highly opportunistic).

At the other extreme of the continuum are the crustose algae. These tend to be comparatively slow-growing, long-lived, persistent species with low surface-area-to-volume ratios, which means a smaller proportion of tissue is exposed to light and nutrients. Many produce calcium carbonate and put a large amount of material into specialized structures, such as complex chambers for reproduction (called conceptacles). Some crusts have adapted to extreme low-light habitats (as deep as 268 meters or in caves and crevices), and others are found on the reef crest, where no delicate fleshy alga could survive the forces of large waves. There are many advantages and disadvantages to any single life form, and certain species have capitalized on this by assuming markedly different forms in their life history (see *Derbesia osterhoutii*), but that is another story and more complex than we will discuss here. The main objective is to be aware that each different form has advantages for survival in the aquatic environment.

HOW TO USE THIS FIELD GUIDE

The first step in using this field guide is to determine if the plant is a seagrass or alga. This is usually evident, as seagrasses have a stiff grasslike appearance. If the specimen is an alga, the next step is to determine to which phylum it belongs, relying on its color. If it is green, yellow-green, or greenish black, see the Chlorophyta section. If brown, ranging from pale beige to yellow-brown to almost black, check the Phaeophyta section. If the specimen is any shade of red, rose, violet, pink, purple, brownish orange, or blackish red, use the Rhodophyta section. For a few species the color can be misleading; red algae in particular include a wide spectrum of shades, especially those growing in the intertidal zone. If the coloration is deceptive or overlapping, look in more than one section.

Next, match the plants to the closest form-group section of the photographs (arranged from delicate to coarse), but remember to consider individual differences. Then read the appropriate descriptions and compare the habitat information and field characteristics to verify your identification.

To aid in understanding the descriptions, we have visually depicted some of the plant structures (Fig. 2), and further explanations of terms appear in the glossary. In many species of algae, the branching pattern is an important characteristic that helps to identify the plant. Figure 3 is a diagrammatic example of some common branching patterns; the glossary at the back of this book has further explanations. Key identifying features, where present, have been indicated by boldface lettering within the species descriptions.

For those individuals who are enthusiastic about underwater photography, a short how-to section is included near the end of this field guide. Also, there are several advanced books available about the marine plants of the Caribbean and surrounding seas. The most technical and complete treatment remains *Marine Algae of the Eastern Tropical and Subtropical Coasts of the Americas,* by W. R. Taylor. That classic work contains detailed descriptions, keys, drawings, geographic ranges, and historical data for all but a few of the plant species we have described in this guide. In brackets at the end of descriptions, we have indicated updates or changes that have occurred since Taylor's book was published in 1960.

Also containing information on and drawings of marine plant life found in the Caribbean and ethnobotany are works by Voss (1976), Woelkerling (1976), Abbott and Dawson (1978), Dawes (1981), Sterrer (1986), Abbott (1984), and others listed in the References section.

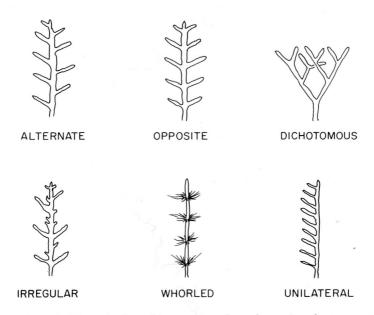

Figure 3. *The major branching patterns shown by marine plants.*

SCIENTIFIC NAMES

Only a few marine plants have common names, and the beginning naturalist may be confused and annoyed by the Latin scientific names. But people use Latin names everyday ("lynx," "alligator," "iris," "geranium," to name a few) and, because everyone is familiar with them, do not notice any difficulty. Latin is used for scientific names because it is a dead language (not used by any group of people for their everyday activities), and consequently it is not subject to change. An organism's scientific name is universal—only one scientific name per type of plant or animal. A specific plant or animal may have many common or popular names, but it will have only one scientific name (see Voss et al., 1983).

An organism's scientific name consists of the generic designation (genus, capitalized) and specific epithet (species, in lower case). One genus may contain many species, all having certain characteristics in common. For example, the genus *Caulerpa* contains many species, but all have runners or stolons (rhizomes) and rootlike rhizoids. The specific epithet sets each species apart from all others within that genus based on its own unique features. For example, *Caulerpa prolifera* has rhizomes (generic character) and rhizoids (generic character), and it is the only species with an upright, solid, sheetlike blade (species character). In botany, the names of the people who originally described the organism and assigned its scientific name (called authors) also are included. Often an author's name will be in parentheses followed by another name. This means that the species was originally described in another genus by the author in parentheses but was subsequently changed to the present genus by the second author.

In some photographs it is impossible to determine precisely what species is pictured. In such cases the genus name is followed by "sp." (as in "*Caulerpa* sp.," to be read "*Caulerpa* species"). This means that the species cannot be determined with the information at hand, and no author will be listed.

Occasionally there are variations within a species that are consistently found; these are referred to as varieties (var.) or forms (f.) and appear in Latin after the generic and specific names. A variety is considered to be a genetic (hereditary) difference, but it does not deviate enough from the original plant to justify a separate species. A form is generally regarded as a variation due to differing environmental conditions; for example, plants of a given species growing in a shady location may consistently look quite distinct from individuals of the same species living in a sunny location. Similarly, a species may show a consistent difference between individuals growing in heavy surf and those found in calm water.

HABITATS FOR COLLECTING CARIBBEAN PLANT LIFE

Reefs

There are four major reef types, each containing a wealth of plant life, and all may be found in the clear waters of the Caribbean. Fringing reefs (Fig. 4), the most common reef type, develop adjacent and parallel to the shoreline. Barrier reefs (Fig. 4) occur some distance from but parallel to the shore and have a well-developed lagoon between the reef and the mainland. The second-largest barrier reef in the world, the Belizean Barrier Reef, is in the Caribbean; it extends for a distance of 240 kilometers and at some points is 40 kilometers from the mainland. The third reef type, an atoll, is a ring of reefs often interspersed with low sandy islands; the outer edges of the reefs rise from great oceanic depths, while the inner ring of reefs and interspersed islands create a shallow lagoon in the center of the atoll. Although abundant in the tropical Pacific, atolls (Fig. 4) such as Glovers Reef, Belize, are not so common in the Caribbean and therefore are not found as frequently as the other reef varieties. The fourth type, patch reefs (Figs. 4 and 5), may be present within the large lagoons of barrier reefs or atolls and occur as small mounds or cup-shaped formations.

The parts of the reef have many names. The terms used here are the most common, but variations may be found in other books or articles. The reef crest (Figs. 6, 7A, and 7B) is that part of the reef that is shallow and sometimes exposed to the air, particularly at low tide. Here, exposure to air and heavy wave action creates too harsh an environment for most marine plants and animals; consequently, organisms found on the crest usually have special characteristics that enable them to survive. Seaward of the crest is the spur-and-groove area (Figs. 6, 7A, and 7B), where raised buttresses are separated by deep channels; this area of high grazing pressure is often heavily affected by hurricanes and other large storms. Seaward of the spur-and-groove system is the fore-reef slope (Fig. 6), which drops steeply into the ocean depths;

SEQUENCE OF ATOLL DEVELOPMENT

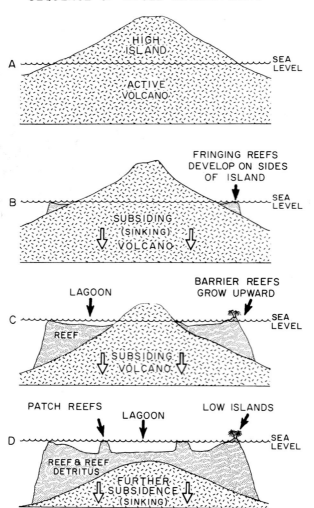

Figure 4. The sequence of fringing reef, barrier reef, and atoll development (first proposed by Charles Darwin).

Figure 5. Aerial view of patch reefs and drift lines of Sargassum *(brown). The crest of the Belizean Barrier Reef is visible in the upper right.*

many delicate corals and algae may live in this habitat beyond the level where most waves or surge can injure them. Landward of the crest is the reef flat (Figs. 6, 7A, and 7B), a relatively shallow calm area behind the protection of the crest. Inward from the reef flat, on barrier reefs and atolls, there is a deep lagoon where certain algae or seagrass beds dominate; patch reefs (Fig. 5) may also be found in these lagoons. Figure 6 depicts the typical reef; however, reef formations are highly variable, and many lack one or more of the parts mentioned here.

Seagrass Beds

Seagrass communities cover vast areas of the shallow-water habitats in the Caribbean (Figs. 7A and 7B). The ribbonlike

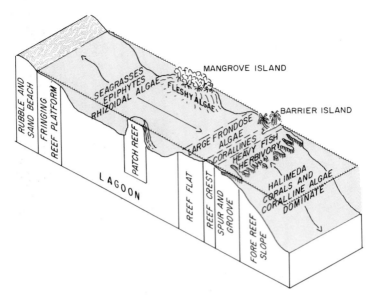

Figure 6. Sectional view through a tropical continental shelf containing characteristic reef and mangrove systems. Dominant plant types are indicated for the various habitats.

Thalassia testudinum and the thinner cylindrical-leaved *Syringodium filiforme* are the dominant plants. Although seagrasses (marine flowering plants) are the most conspicuous organisms within such communities, many species of algae are also found growing on and among these gregarious plants. The majority of seagrasses are tropical in distribution, and the Caribbean holds the second-largest flora in the world (the largest being in the vast Indo-Malaysian region). There are only 45 to 50 seagrass species described worldwide. We have included a spectrum of seagrass species in this handbook so the reader can become familiar with those most common to the Caribbean. The seagrass community, as a whole, is important in stabilizing sediments (owing to the

A.

FORE REEF (SPUR-AND-GROOVE)

REEF CREST

RUBBLE RIDGE

BACK REEF (SAND CHANNELS, SAND PLAINS, AND SEAGRASS)

BACK REEF (ALGAL DOMINATED)

REEF FLAT

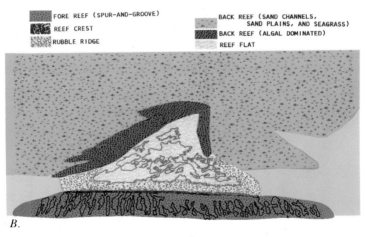

B.

Figure 7. A. Aerial photograph of Looe Key Reef, Monroe County, Florida, showing several major reef habitats. B. Coded diagram of Looe Key Reef delineating the various habitats. Seagrasses dominate the region in brown (see the corresponding dark areas in A, above).

entangled rhizomatous root systems of the seagrasses as well as certain species of sand-dwelling algae) and providing a productive nursery ground for many fishes and invertebrates.

Mangroves

Reef habitats may be the most spectacular and beautiful areas that algae inhabit, but Caribbean mangrove forests (Fig. 8A) also contain a rich and varied algal flora. Tidal mangrove systems occur along sheltered areas on mud bottoms in tropical and, occasionally, subtropical climates. The flowering trees and shrubs that comprise mangrove islands generally grow in the tidal regions of shallow lagoons and estuaries but can also form small islands on offshore barrier reefs or atolls. The trees characteristically have supporting prop roots (Fig. 8B) that extend several meters down into the silty substrate. The prop roots give rise to normal nutritive roots as well as aerial roots that are specialized for gas exchange. The intertidal and subtidal portions of the prop roots support a unique, highly developed, algal vegetation (Fig. 8B). Also, the leafy canopy shades the roots, enabling unusual shade-adapted species to inhabit this nutrient-rich area. Often within the mangrove thickets are small shallow lagoons and channels (Fig. 8A) where stands of mud-dwelling algae occur. To find an even richer habitat for algae, an avid naturalist should locate a mangrove island or shoreline where birds are using the trees as nesting or roosting sites. The bird guano (excrement) fertilizes the already rich environment provided by decomposing leaves and twigs, thereby creating a habitat that can support an incredibly high density of algal vegetation.

A.

B.

Figure 8. A. Aerial view of Twin Cay, Belize, a mangrove island containing rich algal and seagrass floras in the pools and channels. B. Underwater view of mangrove roots and associated marine plants.

Green Algae
(CHLOROPHYTA)

Chlorophyta (from the Greek *chloros,* meaning "green," and *phyton,* meaning "plant"), or green algae, are often present in relatively high numbers in tropical seas. For example, the genera *Ulva* and *Enteromorpha* may proliferate on reefs where nutrients are high, wave shearing forces low, and herbivory reduced. Such genera of green algae are tolerant of stressful conditions, and their presence often indicates freshwater input or pollution. The typical color of plants in the Chlorophyta, a result of the dominant pigment chlorophyll, is some shade of apple or grass green, although certain species range from yellow to black-green. Extremely varied shapes and assemblages are readily seen in the Caribbean, where a striking range of forms reach their most extensive development, including many calcareous species characteristic of tropical waters. Calcified green algae, particularly *Halimeda,* are especially important as the predominant contributors to the production of marine sediments. The sparkling white sand beaches of the Caribbean and many other areas in the world are largely the sun-bleached and eroded calcium-carbonate skeletons of green algae. The deepest occurring, fleshy, upright alga was found attached to bedrock at a depth of 185 meters in the Bahama Islands and is a member of this group.

Ulvaria oxysperma   (Kuetzing) Bliding
ULVACEAE, ULVALES

Translucent, light green, extremely delicate and thin, the sheet-like blades are only a **single layer of cells,** somewhat slick or gelatinous in texture, and they may have flat or somewhat ruffled edges. Plants reach 10 cm tall or larger in protected areas and are attached by a small basal disk that adheres to mangrove roots or other solid substrates in calm waters; also found as soft green tufts in the lower intertidal zone and tide pools to 5 m deep. This species superficially resembles *Ulva* but is much paler in color, thinner, and more delicate. It is limp and tears when handled out of water, much like wet tissue paper. [*Monostroma oxyspermum* in Taylor (1960); see Bliding (1969)]

Ulva lactuca   Linnaeus
ULVACEAE, ULVALES

These bright green, fragile, **thin, ruffled or flat, sheetlike blades** (up to 65 cm in diameter) are attached by a small, inconspicuous holdfast to hard substrates in the intertidal or in shallow, quiet coves and harbors down to 10 m deep. A fast-growing plant, often associated with areas of high nutrients (bird islands), including polluted or recently disturbed habitats where herbivory is low. *Ulva lactuca* can be distinguished from *Ulvaria oxysperma* by its thicker, darker green blades that are composed of two cell layers rather than the one layer found in the latter. In many countries, *Ulva* is eaten in soups, salads, and other dishes.

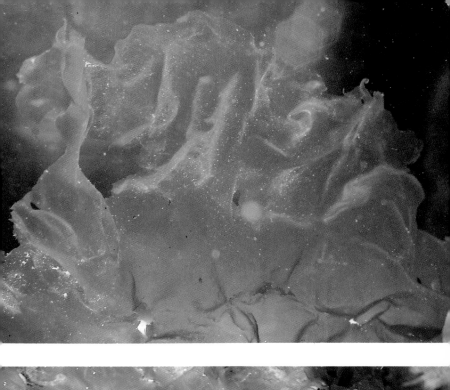

Ulva fasciata   Delile
ULVACEAE, ULVALES

Members of this genus are commonly called sea lettuce. This species is long (up to 50 cm), narrow (3–4 cm), and bright apple green, and it forms **strap-shaped blades** with smooth but ruffled edges. The central portion of the blade is often pale in color, becoming darker toward the margins. Common in the tropics, this plant occurs in shallow waters (high intertidal zone to 10 m deep) on any firm substrate, often in association with high-nutrient areas (e.g., mangroves, bird islands) or near freshwater sources. In Hawaii, this species is mixed with other seaweeds and eaten with raw fish or boiled to form a light soup stock.

Enteromorpha flexuosa   (Roth) J. Agardh
ULVACEAE, ULVALES

Enteromorpha means "intestine-shape," and this alga **resembles thin hollow tubes.** It is light green, unbranched or rarely branched at the base, and reaches a length of 20 cm but is generally smaller. Growing in clusters or tufts at or near the low-tide line, plants are found on mangrove roots, wood, or rock; they often grow epiphytically on other marine plants; common in brackish water of estuaries or around freshwater seeps on sandy shores from the high intertidal zone to 5 m deep. The genus *Enteromorpha* is truly cosmopolitan and may be found in almost any shallow-water brackish or marine environment throughout the world.

Anadyomene saldanhae   Joly and Oliveira
ANADYOMENACEAE, CLADOPHORALES

Rounded, crisp, erect, broad, bright green blades forming loose clusters that can reach 10 cm tall, with faintly visible fan-shaped veins. A magnifying glass or hand lens will reveal the blade's beautiful lacy structure, composed of **randomly arranged cells between the fan-shaped veins.** Attached by a holdfast and short narrow stalk, it grows in cracks and crevices at or just below the low-tide mark to 30 m deep. [see Joly and Oliveira (1968) for description]

Anadyomene stellata   (Wulfen) C. Agardh
ANADYOMENACEAE, CLADOPHORALES

Plants consist of fan-shaped, rounded, crisp, bright green blades up to 10 cm tall. Under a hand lens or magnifying glass, *Anadyomene stellata* blades show a beautiful network of lacy veins. The large **cells between the fan-shaped veins are parallel,** or side by side. Growing in rocky cracks and crevices, plants can be found from the low-tide mark to 30 m deep.

Microdictyon marinum   (Bory) Silva
ANADYOMENACEAE, CLADOPHORALES

Blades resemble a stiff, pale green, **coarse-mesh net.** Individual filaments comprising the net may reach 1 mm in diameter. They are crisp in texture, forming rounded membranes of various shapes up to 6 cm tall and often producing a springy turf. This plant is usually found in areas where the substrate and associated organisms are lightly dusted by fine sediments on rock and corals; it also occurs epiphytically 1–15 m deep. *Microdictyon boergesenii* Setchell, also common in the Caribbean, looks much the same but is composed of smaller filaments.

Polyphysa polyphysoides   (P. and H. Crouan in Maze and Schramm) Schnetter
POLYPHYSACEAE, DASYCLADALES

Very small, inconspicuous plants (5–10 mm tall) composed of a short stalk and topped by only one flat or cup-shaped disk generally 1.5–5.0 mm in diameter. The disk comprises **11–25 rays joined by light lime deposits.** Rays are rounded at the outer tips. These plants are easily confused with *Acetabularia* but are considerably smaller. Found at the bases of larger plants or within shady cracks and crevices, this species is characteristic of shallow (above 5 m) habitats. [*Acetabularia polyphysoides* in Taylor (1960); see also Schnetter (1978)]

Acetabularia calyculus   Quoy and Gaimard
POLYPHYSACEAE, DASYCLADALES

One of the most photogenic of marine algae, consisting of slender stalks (1–3 cm tall) topped with one to several cup-shaped or flattened, lightly calcified, light green disks that are up to 7 mm in diameter and composed of 22–30 rays. **The tips of each ray are generally notched.** Widely distributed in shallow waters (less than 5 m) of mangrove swamps and protected calm areas, small clusters or individuals grow on mangrove roots, shell fragments, or other hard substrates.

Acetabularia crenulata   Lamouroux
POLYPHYSACEAE, DASYCLADALES

"Mermaid's wine-glass" is certainly an appropriate common name for this attractive small (2–8 cm) plant, a favorite of tropical aquarists, it looks much like a parasol of one or more disks or cup-shaped caps above a slender stalk. Moderate calcification (lime encrustation) gives the alga a whitish green color. The cap is composed of 30–80 segments, or rays, and can measure 12–20 mm in diameter. **Each ray of the disk possesses a single tooth or spine on its outermost tip.** Commonly occurring in shallow (1–3 m deep) protected areas, these plants can be found in clusters or as individuals growing on seagrass blades, mangrove roots, rocks, shells, or coral fragments.

Bryopsis pennata var. *secunda*   (Harvey) Collins and Hervey
BRYOPSIDACEAE, CAULERPALES

This variety is an attractive, filamentous, turflike, bushy plant (*Bryopsis* means "mosslike"); it is a shiny dark green color (often with a light blue iridescence) and usually less than 10 cm tall. The branches arise from a tightly interwoven rhizoidal system and resemble delicate feathers with branchlets in one or two rows from the base to the branch tip; **branchlets generally occur only on one side (unilateral) of the upright.** An alga of quiet, shallow (less than 5 m), subtidal waters, it can be found growing on any solid substrate, often on wood such as mangrove roots.

Bryopsis plumosa   (Hudson) C. Agardh
BRYOPSIDACEAE, CAULERPALES

Composed of erect, translucent green (often with a bluish iridescence) tufts of **fine featherlike or plumelike (branchlets in two opposite rows) filamentous uprights,** these plants grow to 20 cm tall. Occurring in the mid- to low-intertidal and shallow subtidal zones to a depth of 5 m, *Bryopsis plumosa* is common in tide pools, sheltered locations, or attached to rock in moderate surf conditions.

Chaetomorpha linum   (O. F. Mueller) Kuetzing
CLADOPHORACEAE, CLADOPHORALES

Plants forming distinctive, large, loosely entangled mats or **mounds up to 1 m tall and 2 m wide** that are yellowish green. *Chaetomorpha* means "hair-shaped," and the cylindrical filaments (each a single row of cells) of this alga are somewhat stiff and curled or twisted, having the texture of coarse steel wool. When examined with a hand lens, the large cells comprising the filaments appear to be joined by dark green bands. Mats often lie free in high-nutrient areas (such as bird islands) in shallow waters (less than 3 m).

Chaetomorpha crassa   (C. Agardh) Kuetzing
CLADOPHORACEAE, CLADOPHORALES

Long filamentous plants resemble bundles of tangled, green, nylon-monofilament fishing line. When examined with a hand lens, a filament can be seen to be composed of a chain of individual **cylindrical cells that are about as long as they are broad.** These plants are usually found growing entangled with other large algae or on mangrove roots in shallow waters.

Chaetomorpha aerea   (Dillwyn) Kuetzing
CLADOPHORACEAE, CLADOPHORALES

Conspicuous only when growing in small, dense clusters, this bright yellow-green alga consists of stiff, unbranched, straight filaments up to 15 cm long. Hanging in small patches on upper-intertidal surf-beaten rocks, *Chaetomorpha aerea* can be found as far north as Maine.

Cladophoropsis macromeres   W. Taylor
HALIMEDACEAE, CAULERPALES

The glossy, light green, turflike, filamentous uprights arise from a basal network of pale or colorless, entangled, fibrous filaments. Upright filaments may have irregular or unilateral branching and spread horizontally over rocky substrates. Growing to 15 cm tall, these plants occur as loose masses or entangle with other algae to form **dense cushionlike clumps** in warm quiet waters, from shallow pools to 5 m deep. A closely related species, *Cladophoropsis membranacea* (C. Agardh) Boergesen, is commonly found adhering tightly to intertidal rocks or wooden substrates as a dense mat of relatively fine filaments.

Derbesia sp.
BRYOPSIDACEAE, CAULERPALES

Forming tight dense turfs of dichotomously branched filaments, this seaweed grows to 5 cm tall and is bright yellow-
green. Species are distinguished microscopically according to
filament size and reproductive structures. These distinctive
turfs are found growing on larger, tougher plants or forming
mats covering mud or sand on mangrove islands to 10 m deep.
Derbesia also has a spherical *Halicystis* stage in its life history (see
page 56).

Cladophora prolifera   (Roth) Kuetzing
CLADOPHORACEAE, CLADOPHORALES

Conspicuous only when occurring in abundance, this seaweed
forms small, stiff balls (up to 8 cm in diameter) composed of
coarse, firm, forked, dark green filaments, but it is normally
present as small tufts (20 cm tall) on rocks in shallow waters
(*Cladophora* means "bearing branches"). If environmental conditions are favorable, unattached *Cladophora* balls can cover vast
areas with accumulations 1–2 m thick in basins of quiet harbors
and inlets ranging from 1 to 10 m deep. *Cladophora* is a large,
complex genus common in both freshwater and marine environments, and because most of the species are so variable in size
and branching, they can be difficult to identify.

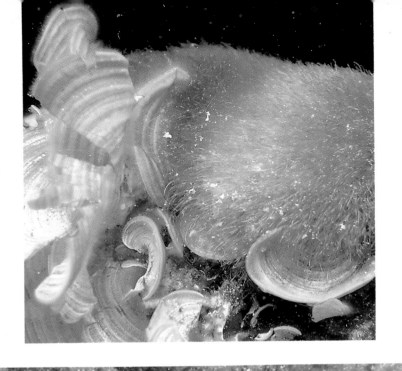

Caulerpa verticillata   J. Agardh
CAULERPACEAE, CAULERPALES

These dark green, fuzzy, felty turfs or mats are composed of small (less than 2 cm in height), delicately whorled uprights connected by a tangle of rhizomes (runners) and rhizoids (roots). Close examination reveals fragile uprights formed by **whorls of fine forked branchlets.** Growing on stable substrates, under overturned rocks, or mangrove roots, or occasionally on peat or mud, this plant is found in shallow protected areas but can occur to a depth of 30 m.

Caulerpa sertularioides   (S. G. Gmelin) Howe
[f. *sertularioides*]
CAULERPACEAE, CAULERPALES

Long, erect, light green, **featherlike blades** (15–20 cm tall) combine to make *Caulerpa sertularioides* a striking plant. Branchlets are cylindrical and needle-shaped, with opposite branching (pinnate) that creates a delicate featherlike appearance. Rhizoids, extending from extensive coarse-branched rhizomes, anchor the plants in shallow (1–10 m) sandy areas or to mangrove roots. *Caulerpa sertularioides* is eaten in salads in some areas of the Philippines.

Caulerpa sertularioides f. *farlowii*   (Weber-van Bosse) Boergesen
CAULERPACEAE, CAULERPALES

Shorter (5–10 cm) than the normal *Caulerpa sertularioides,* this form has small, light to yellowish green, cylindrical, pointed **branchlets loosely arranged all the way around the upright stalk.** Like all other *Caulerpa* species, the upright fronds arise from rhizomes anchored to coral fragments or pebbles by fine tufts of rhizoids; usually found behind the reef crest in shallow waters (less than 8 m) on reef flats.

Caulerpa taxifolia   (Vahl) C. Agardh
CAULERPACEAE, CAULERPALES

This plant resembles *Caulerpa mexicana,* with spreading stolons bearing rootlike rhizoids; pale, grass green, flattened, feather-like uprights (3–15 cm tall) bear small lateral (opposite or pinnate) branchlets, which are narrower than those of *Caulerpa mexicana* and **do not taper into the midrib** but are constricted or pinched at the point of attachment. The branchlets appear like individual leaflets, tapered at both the tip and base; new fronds developing from the midrib of older fronds are common (proliferations). Spreading by rhizomes, this species is found in sand on reef flats in sheltered or moderately wave-exposed areas to 15 m deep.

Caulerpa mexicana   Kuetzing
CAULERPACEAE, CAULERPALES

This species is highly variable in length from small, stubby, 2-cm-tall plants in wave-exposed areas to elongated forms (15–20 cm) in calm, protected habitats. The grass green uprights resemble flattened feathers. The oppositely arranged branchlets, with upward-pointed tips, are relatively wide and taper into the **flat, broad (1–3 mm) midrib.** Plants spread by rhizomes (runners) that are generally attached to small coral fragments or pebbles by fine rhizoids, and occur on sand or mud bottoms in lagoons, mangrove areas, and seagrass beds to 15 m deep.

Caulerpa prolifera   (Forsskal) Lamouroux
CAULERPACEAE, CAULERPALES

Distinctive plants having grass green to dark green, flat, erect, **undivided, oval to elongated blades** (5–10 cm long) with short slender stalks connected by rhizomes from which rhizoids may penetrate deep into sand or mud. The blades often proliferate (hence the name), giving rise to secondary blades off the center portion of the main upright. These are commonly found growing in seagrass beds 1–15 m deep. An easily grown and excellent plant for the marine aquarist.

Caulerpa serrulata   (Forsskal) J. Agardh
CAULERPACEAE, CAULERPALES

The grass green uprights are small (2–4 cm tall), flattened, and often divided dichotomously (a wide Y shape). The stalk is small and inconspicuous, with spiraling twisted blades arising almost directly from the rhizome. **The margins (edges) of the flattened blades possess numerous teeth** (serrulate, or sawlike, as indicated by the name), and like all *Caulerpa*, this species spreads by runners (rhizomes) that contain numerous rootlike rhizoids, typically on sand-covered rock substrates in shallow waters (less than 5 m).

Caulerpa racemosa   (Forsskal) J. Agardh [var. *racemosa*]
CAULERPACEAE, CAULERPALES

One of the most ubiquitous species of this group, forming intertwined mats of whitish yellow rhizomes with upright branches bearing distinctive, grass green, **swollen or beadlike (spherical) tips on the branchlets.** The alga is highly variable in size (1–15 cm tall) and shape, with about a dozen varieties and forms appearing as miniature clusters of grapes. Plants are attached by rootlike rhizoids that cling tightly to the rocks in heavy surf areas and spread by cylindrical rhizomes in the intertidal and shallow subtidal zones on rocky substrates, but they can be encountered down to 20 m deep. Polynesians commonly eat this *Caulerpa*, preferring it raw with freshly grated coconut or coconut milk added.

Caulerpa racemosa var. *peltata*   (Lamouroux) Eubank
CAULERPACEAE, CAULERPALES

Rhizoid-bearing rhizomes give rise to small (1–3 cm tall) **flat-topped (umbrella-shaped) uprights,** each consisting of a small slender "stem" topped with a disklike cap (1.5–8.0 mm in diameter); pale to dark green with a faint bluish sheen. A distinctive variety found in low-light habitats such as shaded mangrove roots, dark crevices, or the underside of ledges in reef habitats, often intermixed with other algal species in turf communities; found in shallow (less than 1 m) to very deep waters (84 m in the clear waters of the Bahamas). [*Caulerpa peltata* in Taylor (1960); see Eubank (1946) for more information]

Caulerpa lanuginosa   J. Agardh
CAULERPACEAE, CAULERPALES

Upright, dark green fronds (about 8–13 cm in height) are connected by rhizomes firmly attached to the solid rock substrate under a thin layer of sand. The rhizomes are also covered by **fine hairs that ensnare and attach to sand grains.** The upright fronds look bushy because of their numerous, short, cylindrical branchlets. Plants are often difficult to locate when growing among *Thalassia* (turtle grass), but are easier to find in open sand-covered areas; found at depths of 1–15 m.

Caulerpa cupressoides (West in Vahl) C. Agardh
[var. *cupressoides*]
CAULERPACEAE, CAULERPALES

Caulerpa cupressoides is highly variable, exhibiting many different forms and varieties; generally tough and stiff, with the upright green portion consisting of **vertical, parallel, columnar branches** connected to one another by runners or rhizomes on the surface of sandy bottoms or slightly buried. Some forms are small and bushy (2–8 cm), whereas others are tall and sparsely branched (15–25 cm). Most varieties have short, knobby branchlets (cylindrical, cone-shaped, or flattened) arranged in vertical rows along the upright axis, whereas other forms have few of the stubby branchlets. Often found growing in abundance on sandy shallow (1–10 m) bottoms with tufts of rhizoids attached to stones and coral fragments or anchored directly in the sand.

Caulerpa cupressoides var. *lycopodium* (J. Agardh)
Weber-van Bosse
CAULERPACEAE, CAULERPALES

This variety of *Caulerpa cupressoides* is quite distinctive, even though immature portions often resemble the typical variety. The mature plant has tall (up to 20 cm), sparingly branched, dull green uprights that are covered by cylindrical branchlets. The uprights arise from an extensive rhizome system (runners) that is anchored in the sand by fine rhizoids attached to coral fragments or small stones. Extensive populations may be found in shallow (less than 5 m deep) sandy or silty areas, occasionally growing on mangrove roots.

Caulerpa paspaloides   (Bory) Greville
CAULERPACEAE, CAULERPALES

Attractive, large, dark green alga, with an unbranched (or occasionally forked) stalk that grows up to 20 cm tall and is topped
with a **forked (divided 2–4 times) crown of finely branched,
soft filaments.** The branchlets are usually arranged in ranks
(rows) of three or four, giving each branch a triangular or
square appearance when viewed from the tip. Thick rhizomes
(runners) and rhizoids anchor the plant on the muddy bottoms
of mangrove channels, lagoons, or shallow (1–12 m) seagrass
communities.

Batophora oerstedii   J. Agardh
DASYCLADACEAE, DASYCLADALES

A fuzzy, bright green, small, cylindrical plant that is most commonly about 3 cm high but can grow up to 10 cm tall. Usually
found forming small, dense clusters. The main axis has tightly
whorled, forked branchlets that give a soft, delicate, cylindrical
appearance to the upright frond. Clusters are found in lagoons,
especially around mangroves in shallow waters (less than 8 m
deep) and are also common in brackish-water habitats (seawater
mixed with much freshwater), often on seashells such as conch.
When a plant becomes reproductive (fertile), the tips of the
branchlets are covered with **small, bright, yellow-green spheres
(several clustered on each branchlet).** The fertile plants in the
group contrast sharply with the attractive sterile (nonfertile)
plants. When this plant is crushed, a distinctive yellow substance
is released.

Dasycladus vermicularis   (Scopoli) Krasser
DASYCLADACEAE, DASYCLADALES

Dasycladus vermicularis is so similar to *Batophora oerstedii* in both appearance and habitat that it is difficult to distinguish the two. Plants are small (3–6 cm tall) and olive green. They look cylindrical and fuzzy, almost spongy, because their whorled branchlets are more tightly packed than *Batophora*'s. Reproductive structures are also bright yellow-green spheres, but only **one sphere (sporangium) appears on each small branchlet.** Found in a variety of habitats, from shallow reefs (less than 8 m deep) and tide pools to lagoons and mangroves, growing on hard substrates such as shells or coral fragments.

Neomeris annulata   Dickie
DASYCLADACEAE, DASYCLADALES

A distinctive, small, erect, cylindrical, fingerlike plant (less than 3 cm tall) that is found as solitary individuals or in dense clusters. Close examination shows that **distinct calcified rings or bands encircle the entire upright.** Whitish toward the base (because of calcification), with a green fuzzy top (formed by whorls of fine hairs), individuals are easily recognized and occur abundantly on shaded mangrove roots, coral fragments, or rocks in shallow, sandy areas, from the intertidal zone to as deep as 30 m.

Siphonocladus tropicus   (P. and H. Crouan in Schramm and Maze) J. Agardh
SIPHONOCLADACEAE, SIPHONOCLADALES

These small (up to 10 cm), pale yellow-green, stiff, bushy tufts consist of many erect branches that have abundant, short lateral branchlets. *Siphonocladus tropicus* often grows beneath ledges or epiphytically on larger algae in calm shallow (above 5 m) habitats.

Ernodesmis verticillata   (Kuetzing) Boergesen
VALONIACEAE, SIPHONOCLADALES

A small, stiff, translucent, yellow-green plant, originating from a single stalk (1–2 cm long) that is topped with a whorl of similar stalks or branches. These in turn are topped with similar branches, eventually forming a small (up to 10 cm tall), bushy, spherical clump with as many as six or more equal levels of branching. When not covered with epiphytes or sediments, this is an attractive plant. Found in shallow (less than 5 m) protected bays on shaded ledges or under larger algae.

Ventricaria ventricosa   (J. Agardh) Olsen and West
VALONIACEAE, SIPHONOCLADALES

Plants look like large glass marbles or small balloons with a bright reflective glare and consist of **large (up to 5 cm in diameter), dark green, thin-walled, oval or spherical, single cells** (one of the largest cells known to science). Individuals tend to be solitary, but on occasion, several may grow together. They are attached to the substrate by minute, hairlike appendages (rhizoids). Specimens are found 1–80 m deep among other algae or in crevices. [*Valonia ventricosa* in Taylor (1960); see Olsen and West (1988)]

Halicystis stage of *Derbesia osterhoutii*   (L. Blinks and A. H. Blinks) Page
BRYOPSIDACEAE, CAULERPALES

Sometimes referred to as "sea bottles" and superficially quite similar to *Ventricaria ventricosa*, these plants also appear as small (up to 3 cm diameter), **shiny green balloons or large glass marbles.** They occur singly or in clusters, each consisting of one giant cell attached to the substrate by an inconspicuous, stalk-like, basal holdfast. *Derbesia osterhoutii* differs from *Ventricaria ventricosa* in several ways: it has a rich, pale bottle green color (*Ventricaria ventricosa* has an iridescent luster), it floats in seawater (*Ventricaria* is denser and sinks), it is pliable to the touch (*Ventricaria* is harder and firmer), and *Derbesia osterhoutii* nearly always grows on crustose coralline algae such as *Sporolithon* and *Hydrolithon*. Plants occur in deep water (to at least 18 m) or shallower in shaded cracks and crevices. Laboratory studies showed that spherical forms such as the *Halicystis* stage pictured here reproduce to give rise to filamentous plants, such as the *Derbesia* on page 36. [*Halicystis osterhoutii* in Taylor (1960); see Page (1970) for laboratory experimentation]

Valonia aegagropila   C. Agardh
VALONIACEAE, SIPHONOCLADALES

Composed of **tightly packed, straight, elongated, or bubble-shaped branchlets,** these plants form olive green oval cushions, tufts, or rounded balls 4–20 cm in diameter. The younger individuals are loosely attached, and older plants are often free living. Each branchlet consists of a single cell, 1–3 mm in diameter and approximately 5–10 mm long. Attached to any hard substrate or lying free (unattached, rolling on the bottom), this alga is found from just below the low-tide mark to 5 m deep in the protected waters of the back reef or lagoon.

Valonia utricularis   (Roth) C. Agardh
VALONIACEAE, SIPHONOCLADALES

This olive green species characteristically **creeps or spreads along the substrate,** with occasional offshoots becoming erect (up to 5 cm tall). Smaller than *Valonia macrophysa* and not forming massive mats, this species consists of bubblelike, branched filaments composed of relatively large, cylindrical cells up to 2 cm long but only **3 mm in diameter.** Attached to hard substrates in cracks, crevices, or other protected areas, these plants are mainly found on shallow (5 m) reef flats.

Valonia macrophysa   Kuetzing
VALONIACEAE, SIPHONOCLADALES

Plants are made up of large, irregular or club-shaped, dark olive green, bubblelike cells 5–20 mm in diameter. Often found growing in **large, crowded masses tightly adhering to the substrate,** this species commonly occurs in shaded or darkened areas such as rock crevices or under ledges and is generally found in shallow waters (less than 5 m). *Valonia macrophysa* is often difficult to locate because it is normally overgrown by other species.

Dictyosphaeria ocellata   (Howe) Olsen-Stojkovich
VALONIACEAE, SIPHONOCLADALES

Plants form dense, **firm crusts of small, angular, green, lens-shaped cells** (cells are easily visible to the unaided eye). These bubbly crusts grow to 10–20 cm broad and up to 3 cm thick. Elongated, rhizoidlike cells that are also rather firm attach them tightly to hard substrates: on intertidal rocks exposed to moderate wave action and occasionally on mangrove roots. [*Valonia ocellata* in Taylor (1960); see Olsen-Stojkovich (1985) for an update]

Dictyosphaeria cavernosa   (Forsskal) Boergesen
VALONIACEAE, SIPHONOCLADALES

Commonly called green bubble weed, this light green alga is sacklike, hollow, and spherical when young or irregularly lobed and open when old (because of ruptures). It may reach more than 10 cm in diameter, and the **sphere wall is composed of one layer of large angular or polygonal cells** that are clearly visible with the unaided eye, appearing much like an irregular honeycomb when examined closely. It is lightly attached to the substrate by rhizoids and occurs 1–30 m deep on rocks and dead coral heads, often forming extensive mats. Like *Cladophora prolifera*, *Dictyosphaeria cavernosa* can become an over abundant nuisance under conditions of high nutrients.

Codium isthmocladum   Vickers
CODIACEAE, CAULERPALES

This plant, also called dead man's fingers, is bushy, erect, dark green, and up to 20 cm tall, with fine hairs usually arising from the soft, spongy, smooth surface. **Branches are cylindrical** and rarely flattened where they divide; dichotomous branching occurs near the tips, but irregular branching is common in other parts of the plant. This seaweed often forms compact hemispherical clumps attached by a single disklike holdfast in shallow waters (1–10 m) on hard substrates.

Codium decorticatum   (Woodward) Howe
CODIACEAE, CAULERPALES

Large (up to 1 m, but more commonly 25–40 cm tall), deep yellow-green, and bushy, *Codium decorticatum* often forms a hemispherical or dome-shaped clump. Branches are cylindrical but **flattened at each dichotomous fork.** Commonly observed in shallow waters (above 15 m), these plants attach by a single holdfast to rock or other hard substrates. The word *Codium* is Greek and means "skin of an animal," referring to the soft, spongy texture.

Codium repens   Vickers
CODIACEAE, CAULERPALES

A dark dull green alga adhering tightly to rocks, *Codium repens* forms **mats or patches (reaching 40 cm in diameter) of flattened, widely dichotomous to irregular branches** that appear to grow over or cling to rock. It is thick, tough, and rubbery in texture; tips of branches are blunt and rounded. Most commonly occurring under ledges on the reef crest, this species is found from the lower-intertidal zone to a depth of 10 m.

Codium intertextum   Collins and Hervey
CODIACEAE, CAULERPALES

This alga forms dark green, firm but spongy, slick, **smooth, creeping mats that tightly adhere to rocks,** sometimes forming overlapping broad-lobed or cushion-shaped plants up to 5 cm thick. Fine hairs cover the surface, creating a halo effect when submerged. *Codium intertextum* often dominates to form a distinct zone near the low-tide mark, but is also found as deep as 20 m.

Rhipilia tomentosa   Kuetzing
UDOTEACEAE, CAULERPALES

Plant dark green, to 5 cm tall, **thick and spongy, with a flattened, fan-shaped blade** atop a short stalk. Found as solitary individuals or in small clumps, this species has been reported mostly from deep waters; however, we have found it as shallow as 2 m deep in the Florida Keys.

Avrainvillea rawsonii   (Dickie) Howe
HALIMEDACEAE, CAULERPALES

This species is dull, dark green, almost black, consisting of distinctive **fingerlike or conical projections,** possessing a spongy texture. These unique fingerlike uprights may grow to 15 cm tall and emerge from a massive basal structure of tangled rhizoids. Plants are found in intertidal or very shallow (less than 1 m) subtidal zones on sand-covered substrates, often in seagrass communities.

Avrainvillea nigricans   Decaisne
HALIMEDACEAE, CAULERPALES

Dark greenish brown to blackish plants, with flat, broad, spongy, paddle-shaped blades on short stalks up to 20 cm tall. The blades have a soft **suede appearance and texture,** are somewhat thick and tough (but flexible), and are composed of fine interwoven filaments not easily seen. A basal mass of rootlike rhizoids anchors this species in fine, mud sediments in shallow, protected areas (to a depth of 30 m), often near mangrove trees. Although frequently looking almost identical to *Avrainvillea longicaulis* externally, *Avrainvillea nigricans* differs microscopically in its anatomical construction.

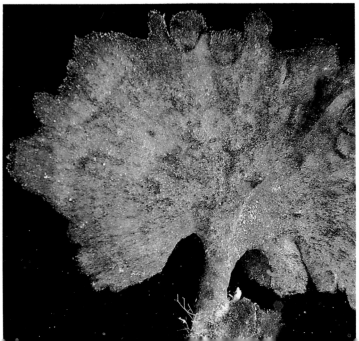

Avrainvillea longicaulis   (Kuetzing) Murray and Boodle
HALIMEDACEAE, CAULERPALES

Plants to 20 cm tall consist of dark green to blackish, broad to oblong, paddle-shaped, flattened blades that are composed of many microscopic, branched, densely packed filaments. The stemlike cylindrical stipe arises from a massive, bulblike, basal holdfast consisting of closely packed rhizoids embedded in sandy or muddy substrates. This species is most commonly found in shallow protected waters of mangrove lagoons but also occurs as deep as 30 m.

Avrainvillea elliottii   A. and E. S. Gepp
HALIMEDACEAE, CAULERPALES

A distinctive species having a broad fan-shaped top (up to 10 cm wide) with the **lower margin quite straight** (most other *Avrainvillea* species have a rounded or flared lower margin), angling sharply into the slightly flattened stalk. Brownish green with light zonations apparent, the moderately thick blades can grow to 13 cm tall. Individuals are found on sand or gravel bottoms to a depth of 10 m. This species has been reported only from the Lesser Antilles and the Grenadines, where it is relatively common.

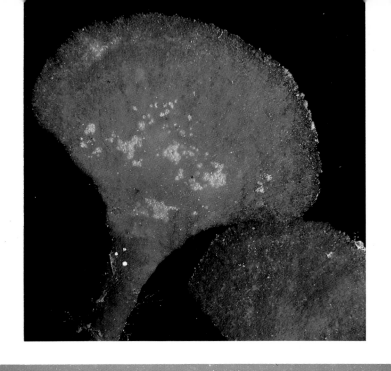

Avrainvillea asarifolia   Boergesen
HALIMEDACEAE, CAULERPALES

Dull, dark, grayish green alga that grows to 30 cm tall. The paddle-shaped blades can be up to 20 cm broad and 15 cm tall. The blade has a smooth edge but is **deeply lobed where it joins the stalk,** thus forming an inverted heart shape. Blades generally have a smooth, firm texture and faint, concentric zonation. The stalk is long (6–16 cm) and relatively slender compared with other species in this genus. Individuals are found in moderately deep (to 35 m) sandy areas.

Cladocephalus luteofuscus   (P. and H. Crouan) Boergesen
UDOTEACEAE, CAULERPALES

Dark yellow-green plants, to 30 cm tall, consisting of a flattened blade above a long stipe that arises from a large clump of matted rootlike rhizoids. The main blade often has **additional blades radiating from the sides,** a feature seldom found in the closely related, similar genus *Avrainvillea.* These two genera differ structurally; *Cladocephalus* has a microscopic outer layer (cortex), composed of a meshwork of fine clear filaments, that makes the plants tougher textured than *Avrainvillea.* This species is commonly found in shallow mangrove lagoons, often growing with *Avrainvillea longicaulis.*

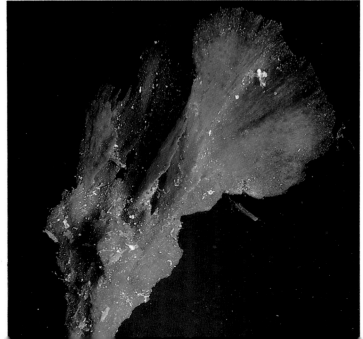

Udotea cyathiformis   Decaisne
HALIMEDACEAE, CAULERPALES

Attractive plants reaching a height of 15 cm and consisting of a whitish green, lightly calcified funnel or cup-shaped blade attached to a small, single stalk. Ranks among the more delicate members of this genus, thin and papery with a **smooth surface to the cup-shaped cap,** which splits easily. This species occurs 1–20 m deep, anchored by elongated rhizoids in sand or finer sediments.

Udotea flabellum   (Ellis and Solander) Lamouroux
HALIMEDACEAE, CAULERPALES

Sometimes called mermaid's fan, this broad, firm, fan-shaped, green blade rests atop a short stalk. Individual plants may attain a height of 20 cm. The stiff blade has distinctive concentric lines or zones with a smooth surface (filaments cannot be seen with the naked eye). It is flexible but also tough and leathery; the edges can be straight, ruffled, or undulated. Light deposits of calcium carbonate (lime encrustations) are present within the blade and stipe. Held upright in sand by a dense bulblike mass of rhizoids, these plants are common in sand patches and seagrass beds over a broad depth range (1–30 m).

Udotea occidentalis   A. and E. S. Gepp
HALIMEDACEAE, CAULERPALES

A **light gray-green, fan-shaped blade** sitting atop a short stalk, the entire plant may reach a height of 8–10 cm. The blade and stalk are well calcified, with the upper edges greener and showing faint concentric zones. The main blade typically proliferates (many smaller blades grow from the central portion). Individuals can be found anchored in sandy substrates by a dense bulblike mass of rhizoids to 40 m deep.

Udotea wilsonii   A. and E. S. Gepp and Howe in A. and E. S. Gepp
HALIMEDACEAE, CAULERPALES

The resemblance to the spokes of a wagon wheel is distinctive when observed from above. *Udotea wilsonii* is slightly calcified and green, may grow up to 13 cm tall, and consists of a short stalk topped by a somewhat spherical crown composed of **thin, platelike fins radiating** from the upper half of the central stalk. Individuals are occasionally found in sandy areas among seagrasses in the shallow (above 10 m) subtidal region.

Rhipocephalus phoenix   (Ellis and Solander) Kuetzing
[f. *phoenix*]
HALIMEDACEAE, CAULERPALES

A stalked, dark green plant up to 10 cm tall, **consisting of concentric flattened platelets** (resembling a pine cone) that form a symmetrical, tightly packed (arranged close to the main stalk) cylindrical cap. Generally occurring in shallow waters, these plants often live among seagrasses; however, some specimens are found to depths of 40 m on rock or sand. This species has two other recognized forms that are distinctive and commonly found on reef flats.

Rhipocephalus phoenix f. *brevifolius*   A. and E. S. Gepp
HALIMEDACEAE, CAULERPALES

Similar to *Rhipocephalus phoenix* [f. *phoenix*], although the platelets of this form are not so densely arranged on the stalk, overlapping in shinglelike series along the **long and narrow cap.** The dark green cap may reach a length of 15 cm. Populations are generally found in sandy or silty areas from shallow waters to 40 m in depth.

Rhipocephalus phoenix f. *longifolius*   A. and E. S. Gepp
HALIMEDACEAE, CAULERPALES

Also similar to *Rhipocephalus phoenix* [f. *phoenix*]; however, this form has a broader, proportionately shorter (3–4 cm long and about as broad), dark green head; the entire plant can grow to 15 cm tall. **Platelets are arranged loosely** around the main stalk and are not as compact as in other forms, giving a rough or unkempt appearance. It grows on rocky or sandy substrates to a depth of 40 m.

Penicillus capitatus   Lamarck
HALIMEDACEAE, CAULERPALES

Mermaid's shaving brush, as its common name implies, resembles a shaving brush. Its faded green, neat, well-defined, bristly, lightly calcified cap sits atop a 5–10 cm stalk that is more heavily calcified. The cap, **normally as long as it is broad,** is composed of relatively slender, dichotomously branched filaments. A characteristic and common plant of calm shallow (above 12 m) lagoons and bays with mud or sandy bottoms; individuals are widely scattered, often intermixed with seagrasses or among mangrove roots.

Penicillus dumetosus   (Lamouroux) Blainville
HALIMEDACEAE, CAULERPALES

Penicillus dumetosus has **bristlelike filaments that taper into the somewhat flattened, relatively short stalk.** The bright green, densely compacted, coarse cap is lightly calcified, and the entire plant can reach a height of 15 cm. This alga can be found firmly anchored in sandy, shallow (2–15 m), well-protected areas by a matrix of rhizoids that intertwine with sand grains.

Penicillus pyriformis   A. and E. S. Gepp
HALIMEDACEAE, CAULERPALES

Rather than having a rounded cap, as in *Penicillus dumetosus,* these individuals have a **flat top with stiffer bristles** that are more heavily calcified and grow to 12 cm tall. The flat-topped tuft is gray-green, dense, and stiff, and it tapers into the stalk. Held firmly in the sand by compacted rhizoids forming a bulbous anchor, these plants occur in sandy, shallow, protected areas 10–30 m deep.

Chamaedoris peniculum   (Solander) Lamouroux
SIPHONOCLADACEAE, SIPHONOCLADALES

The bright, yellow-green, funnel- or oval-shaped cap is composed of coarse, spongy filaments and sits atop a fragile, cylindrical, slightly calcified, **erect stalk (up to 10 cm tall) that is
divided into small segments by rings** (annular constrictions).
This species is not common but is occasionally found forming
small clusters in shallow waters to 10 m deep. The small purple
to rose calcified crusts typically found on the stalk of *Chamaedoris*
are *Fosliella chamaedoris* (Foslie and Howe) Howe, named for the
plant on which it most often occurs.

Cymopolia barbata   (Linnaeus) Lamouroux
DASYCLADACEAE, DASYCLADALES

A unique and striking plant with branches that consist of cylindrical, heavily calcified, **white segments containing tufts of fine,
soft, short, bright green filaments protruding from the tips;**
5–30 cm tall. Branching is irregularly dichotomous. Plants occur in warm shallow waters (less than 5 m) on rocks or coral
fragments. In some quiet bays or harbors, *Cymopolia barbata* is so
common in the shallows that it forms a dense zone adjacent to
the shoreline.

Halimeda lacrimosa   Howe [var. *lacrimosa*]
HALIMEDACEAE, CAULERPALES

This species differs from other members of this genus in having calcified **segments that are spherical or tear-shaped beads** forming a chain that resembles a pearl necklace. Plants are small (2–5 cm tall when shallow, but up to 25 cm long in deeper waters), fragile, straggling, gray-green to whitish with no distinct stalk. Branching varies from irregular to dichotomous. This variety grows on rocks 1–61 m deep and is most commonly found in the Bahamas, Florida Keys, and Cuba. Another variety, *Halimeda lacrimosa* var. *globosa* Dawes and Humm in Dawes, has larger segments (to 9 mm in diameter) and has been found only in the deep waters of the Bahamas (37–91 m). [see Dawes and Humm (1969) for more detail]

Halimeda goreaui   W. Taylor
HALIMEDACEAE, CAULERPALES

The best characteristic for easy identification is the **size of the relatively small, flat, three-lobed segment** (to 4 mm in diameter) and the arrangement of the segments on the outer sections of the branches in a straight line with little or no branching. Plants can grow up to 13 cm long, and the flat, green, brittle, moderately calcified segments have three lobes on the top edge and are slightly ribbed. Most branching originates from a single row of segments attached to hard substrates by a strong holdfast. *Halimeda goreaui* is found hanging in shaded areas under crevices or in deep waters (to 80 m). This photograph also depicts the larger species *Halimeda copiosa*. [see Taylor (1962) for more information]

Halimeda copiosa   Goreau and Graham
HALIMEDACEAE, CAULERPALES

Composed of calcified, disk-shaped or flat, squarish (on older plants) segments that are bright green on the upper surface and often white on the lower surface and up to 1.5 cm wide by 1.0 cm long. This species sometimes forms **long chains** (normally 10–15 cm long, but have been found up to 40 cm long in the Bahamas). Branching initially is in one plane; however, this feature is obscured with age, and branching seldom occurs near the outer tips. When exposed to air or bright light, the entire plant turns white (the chloroplasts, green pigmented structures, migrate internally away from the bright light). Attached to hard substrates by an inconspicuous holdfast, *Halimeda copiosa* is a plant of low-light habitats growing in relatively deep waters (15–40 m) on the vertical sides of rocks or under overhangs and ledges, usually hanging downward from its holdfast. [see Goreau and Graham (1967)]

Halimeda tuna   (Ellis and Solander) Lamouroux [f. *tuna*]
HALIMEDACEAE, CAULERPALES

The lightly calcified, dark green, thin, disk-shaped segments are connected by flexible joints; segments are relatively large (up to 1.5 cm in diameter), with smooth to slightly notched edges and no ribs. The mature plant is 10–25 cm tall. Initially, branching is in one plane, but the trait is obscured with age. Basal segments form a distinct stalk on older individuals. *Halimeda tuna* is common on hard substrates but inconspicuous because it grows as scattered individual plants 10–70 m deep.

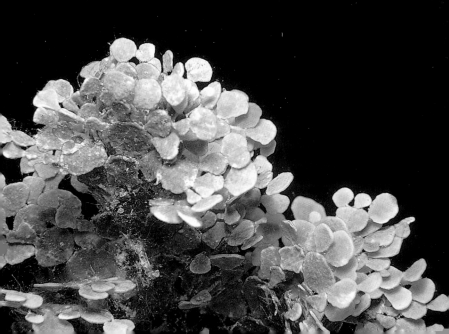

Halimeda tuna f. *platydisca*   (Decaisne) Barton
HALIMEDACEAE, CAULERPALES

Unlike the standard *Halimeda tuna* [f. *tuna*], which has a distinct
stalk, this form generally has no distinct stalk, and it has larger
(2 cm or more in diameter), thinner, disklike segments connected by flexible joints. Dark green in color with very light
calcification, the segments are often somewhat flexible (brittle
when dried). Larger outer segments generally have a rippled
edge. Individuals are found scattered on hard substrates in waters from 10–50 m deep.

Halimeda discoidea   Decaisne
HALIMEDACEAE, CAULERPALES

As in all *Halimeda,* the calcified segments of this species are
connected by noncalcified flexible joints. The segments are
lightly calcified, bright green to white, ribless, disk-shaped, flat,
slick, and soft; segments are the largest known for this genus
(2–4 cm in width). Individual plants grow to 20 cm tall with few
branches; initially branching is in one plane, a feature that is
obscured with age. *Halimeda discoidea* has a small but obvious
holdfast that attaches to shells and hard surfaces in sand
throughout shallow to moderately deep waters (1–80 m).

Halimeda incrassata   (Ellis) Lamouroux
HALIMEDACEAE, CAULERPALES

These erect, light green plants are composed of heavily calcified, hard, brittle, disklike, **distinctly ribbed** and/or somewhat
lobed segments, connected by noncalcified flexible joints. They
grow to 25 cm tall. The lower stalk, which may have several
forked branches, consists of fused (inflexible), often cylindrical
segments. Initial branching is in one plane, a trait that is obscured with age. This species is often associated with seagrasses
on shallow (less than 12 m) sand flats, where it is anchored by a
single bulblike mass of rhizoids tightly gripping the sand.

Halimeda monile   (Ellis and Solander) Lamouroux
HALIMEDACEAE, CAULERPALES

Halimeda monile consists of hard, calcified, flattened, dark green
segments connected by flexible joints; it reaches 20 cm tall. The
segments are partially flattened as in *Halimeda incrassata*, with
three lobes in the lower portions, but **the outer tips of the
branches are almost cylindrical.** Initial branching is in one
plane, but the feature is obscured with age. It often occurs 1–
12 m deep on sand flats among seagrasses, and as in *Halimeda
incrassata*, the rhizoids bind firmly with sand to form a substantial bulblike anchor.

Halimeda opuntia   (Linnaeus) Lamouroux
HALIMEDACEAE, CAULERPALES

Whitish green to dark green, this alga consists of hard, disklike, calcified segments connected by flexible joints; grows 10–20 cm tall; and spreads laterally with **dense random branching to form extensive clumps or mounds with many points of attachment.** Segments of *Halimeda opuntia* are usually three-lobed or rounded with three radiating ribs visible on the surface. It grows on sand or gravel and in *Thalassia* beds down to 25 m deep. The small calcified segments break up to form sand particles making *Halimeda*, and this species in particular, the major producer of carbonate sands and sediments throughout the Caribbean.

Brown Algae
(PHAEOPHYTA)

Members of the Phaeophyta (from the Greek *phaios,* meaning "brown") are almost exclusively marine and attain their greatest abundance, size, and diversity in cold temperate waters (e.g., the Chilean and Californian kelp forests). Tropical waters have comparatively few species of brown algae, although genera such as *Sargassum* and *Turbinaria* can be dominant in some areas, forming small-scale forests up to several meters in height that provide habitat and shelter for many other organisms. *Sargassum* is also unique among seaweed genera in that it contains totally free-floating species with no requirement for attachment to the bottom, as in the Sargasso Sea. *Sargassum fluitans* and *Sargassum natans* are the two dominant plants of the Sargasso Sea, which was named by fifteenth-century Portuguese sailors who thought the air bladders present on these algae resembled *Sarga,* a kind of grape. Drift lines of *Sargassum fluitans* and *Sargassum natans* can often be seen on the ocean's surface (Fig. 5), and both species are frequently found in beach drift. The colors of brown algae (predominantly due to the brown pigment fucoxanthin) cover a spectrum from pale beige to yellow-brown to almost black. In tropical seas, they range in size from microscopic filaments to *Sargassum* species several meters in length.

Dictyota linearis   (C. Agardh) Greville
DICTYOTACEAE, DICTYOTALES

One of the smaller (up to 12 cm), finely branched species in this genus, it differs in having dichotomous **branches of narrow, uniform width (1 mm or less)** throughout the plant. This species often forms brown, bushy, tangled clumps from near the low-tide line to a depth of 20 m and can be attached to any appropriate substrate, including rocks or other algae.

Dictyota divaricata   Lamouroux
DICTYOTACEAE, DICTYOTALES

One of the narrowest and smallest species (under 10 cm high) of *Dictyota;* the branches in the lower portion are up to 4 mm wide, and **with each division, branches become progressively narrower** upward toward the tips, which become very fine. Brown in color, the plants often have a greenish surface iridescence (luster) on the widely dichotomous branches (90° to 120° angles). Characteristically found growing in a tangled mass, individuals attach to rocks or shell fragments in sandy, shallow waters but are also reported as deep as 70 m.

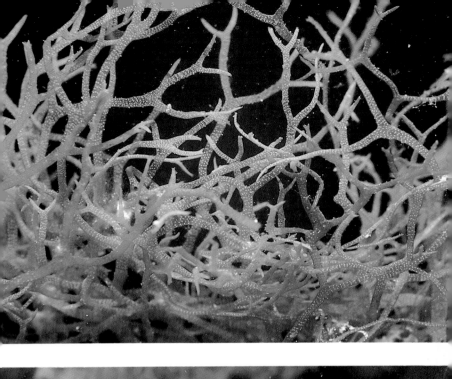

Dictyota bartayresii   Lamouroux
DICTYOTACEAE, DICTYOTALES

Dictyota bartayresii usually forms small, dense clumps (5–10 cm), often tightly adhering to or appearing to creep over rocks. Color is light brown, often with a **brilliant blue iridescence (luster)**. Plants are delicate and tear easily. Branches are dichotomous (in a Y pattern) with broad rounded tips lacking a midrib. *Dictyota bartayresii* is usually found attached to rocks in protected areas, mostly in shallow waters, but it can occur in deep waters (to 40 m).

Dictyota cervicornis   Kuetzing
DICTYOTACEAE, DICTYOTALES

Plants are composed of thin, flat, olive brown blades up to 25 cm long and of various widths. They often grow in a twisted or spiral manner and do not branch in an equal dichotomy; one branch at a point of dichotomy will be much smaller, and the other, larger branch will continue to divide. This cervicorn branching (hence the name) is responsible for the uneven growth pattern. Individuals attach to rocks, shell fragments, or larger plants in sandy shallow areas (1–15 m).

Dictyota jamaicensis   W. Taylor
DICTYOTACEAE, DICTYOTALES

This bushy yellow-brown alga grows to 20 cm tall. The strap-shaped, often twisted branches divide dichotomously at wide angles in the lower portions of the plant but form narrow, almost parallel angles in the upper regions. The blade margins possess **small teeth** at irregular intervals. A small, inconspicuous, fibrous holdfast attaches the alga to hard substrates in moderately shallow waters (1–15 m deep).

Dictyota mertensii   (Martius) Kuetzing
DICTYOTACEAE, DICTYOTALES

This handsome brown plant, often with an iridescent blue-green sheen to the large fluffy fronds, is large (to 20 cm tall) and bushy with a distinctively broad main axis. The lateral **alternate branches** contain numerous short, spurlike branchlets. Frequently found in moderately wave-exposed areas where fish grazing is minimal, this species grows from the intertidal zone to 15 m deep.

Dictyota ciliolata   Kuetzing
DICTYOTACEAE, DICTYOTALES

Growing to 15 cm tall, this species has **distinctive scattered teeth** along the edges of the strap-shaped blades. Light brown blades are irregularly dichotomously branched and spirally twisted in bushier individuals. Plants attach to rocks or other hard substrates by individual small fibrous holdfasts in moderately wave-exposed habitats 1–10 m deep.

Dictyota dichotoma   (Hudson) Lamouroux
DICTYOTACEAE, DICTYOTALES

A bushy, yellow-brown to dark brown plant with thin, strap-shaped blades that are about 15–35 cm long. **Branches have smooth margins and are regularly dichotomous** (dividing into two equal parts) with blunt tips. Distributed worldwide in tropical waters 1–30 m deep, plants are attached on small rocks or coral fragments in sandy areas by an inconspicuous, rhizoidal holdfast. For those who dive, this and other *Dictyota* species are excellent for keeping face masks from fogging. Scrub both the inside and outside of the mask with a handful of reasonably clean *Dictyota,* making sure no sand is attached, then rinse with seawater.

Rosenvingea intricata   (J. Agardh) Boergesen
SCYTOSIPHONACEAE, SCYTOSIPHONALES

Superficially resembles a thick *Dictyota;* a closer look, however, reveals it is not flat but consists of **hollow,** golden to olive brown, branched cylinders. Plants form either entangled mats or bushy individual clumps that can grow to 40 cm in height. The main branches are wide but gradually taper toward the somewhat **pointed tips.** This species is attached by numerous small inconspicuous holdfasts to various hard substrates or occasionally seagrasses; it is found from lower intertidal waters to 35 m deep, from sheltered to wave-exposed habitats.

Rosenvingea sanctae-crucis   Boergesen
SCYTOSIPHONACEAE, SCYTOSIPHONALES

Similar to *Rosenvingea intricata*, but with **blunt branch tips.** The golden to dark brown, **hollow** branches usually divide dichotomously, although some may branch irregularly (like antlers). Plants are usually about 20 cm high (up to 2 mm in diameter) but have been reported to 40 cm. They grow attached to stones and other hard substrates in sheltered, shallow, subtidal areas to 15 m deep. [may be the same as *Rosenvingea orientalis* (J. Agardh) Boergesen; see Earle (1969) for details]

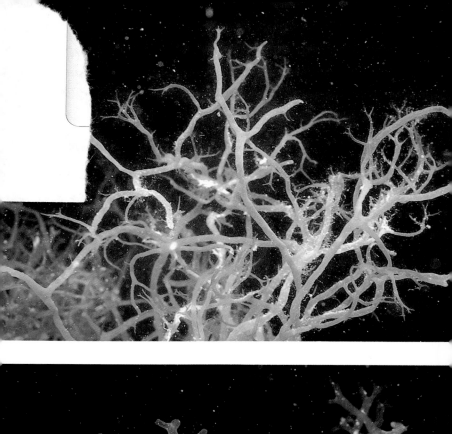
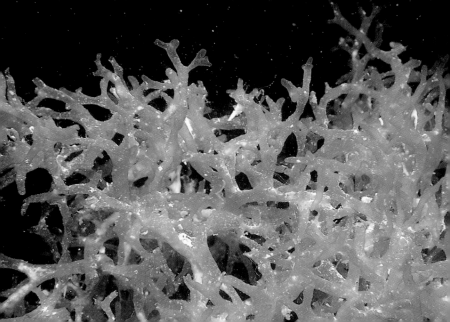

Dictyopteris delicatula   Lamouroux
DICTYOTACEAE, DICTYOTALES

May be confused with a small *Dictyota*, but *Dictyopteris delicatula* differs in having a **very fine midrib** (center vein) throughout. The color is generally a darker golden brown than *Dictyota*, and plants reach a height of about 8 cm. The straplike blades can be either erect or spreading and entangled with dichotomous branching. This alga attaches to rocks or coral fragments from the low-tide level to 12 m deep.

Dictyopteris jamaicensis   W. Taylor
DICTYOTACEAE, DICTYOTALES

Plants are similar to *Dictyopteris delicatula* but much larger (up to 15 cm tall), with dichotomous branching only at the tips; all of the main branching is alternate. Thin, brown, strap-shaped blades (7–11 mm wide) possess a **readily apparent midrib** (center vein), and the blade edges are usually smooth or occasionally contain minute teeth. In the lower, worn-away portions of the plant, the older midrib forms the stipe (stem). Individuals are generally attached to hard substrates by a distinct holdfast and are occasionally found in shallow habitats (1–10 m deep).

Dictyopteris justii   Lamouroux
DICTYOTACEAE, DICTYOTALES

Large (up to 40 cm tall and 2–8 cm broad), fleshy, dichotomously branched, dark yellow to olive brown plant with firm, strap-shaped blades that possess a **heavy, yellow-brown midrib** (middle vein). Blade tips may be indented or blunt, and the edges are usually slightly ruffled. This species is found on hard substrates in moderately wave-exposed areas from the intertidal zone to 30 m deep.

Dictyopteris jolyana   Oliveira and Furtado
DICTYOTACEAE, DICTYOTALES

Large plants (to 50 cm tall) that can look much like *Dictyopteris justii*. However, the blades (to 6 cm wide) are thicker and considerably tougher; when they are examined closely, short, fine hairs can be seen covering the surface. The blades are yellowish olive brown (often with darker brown spots), and the edges are ruffled; the **tips are blunt and indented or notched**. Distinctly dichotomous branching is present only near the base, where the short stipe (stem) is attached to rocks or other hard substrates by small fibrous rhizoids. These plants grow in moderately wave-exposed areas from the intertidal region to 30 m deep. [see Oliveira Filho and Furtado (1978) for further information]

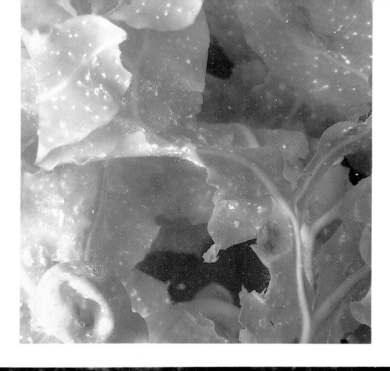

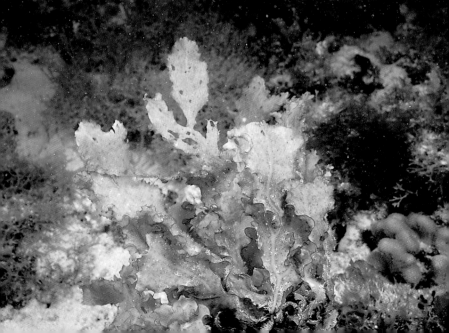

Stypopodium zonale   (Lamouroux) Papenfuss
DICTYOTACEAE, DICTYOTALES

An attractive, large, showy plant that grows up to 40 cm tall and forms clumps of flat blades. Color is yellow brown to dark brown, often with an iridescent green-brown sheen. **The fan-shaped, iridescent blades lack an in-rolled margin** and generally become irregular from differential (uneven) growth or splitting. The surfaces of the blades have zones or concentric bands formed by rows of microscopic colorless hairs. Common in shallow waters (less than 1 m), plants can also be found as deep as 80 m. Individuals attach to rock fragments in sandy areas or on solid rocky substrates by a strong holdfast. This species contains potent chemicals that discourage grazing by fishes.

Padina gymnospora   (Kuetzing) Sonder
DICTYOTACEAE, DICTYOTALES

Like *Padina sanctae-crucis*, the fan-shaped blades (10–15 cm tall) of this alga form leafy clusters. Brown or tan with concentric lines formed by rows of microscopic hairs and reproductive structures, the blades are **lightly calcified or even uncalcified.** This genus is always identifiable by the in-rolled outer margin of the fan-shaped blade. It is found in sheltered or moderately exposed intertidal areas to 14 m deep, usually attached to rocks, corals, or mangrove roots. [*Padina vickersiae* in Taylor (1960); see Allender and Kraft (1983) for details]

Padina sanctae-crucis   Boergesen
DICTYOTACEAE, DICTYOTALES

Light brown to chalky white (the white color is from thin deposits of chalky calcium carbonate on the upper blade surface), species of *Padina* represent the only brown algae that calcify. The blades are thin and fanlike or cup-shaped (to 15 cm tall) with in-rolled margins. The concentric zones of the blade surface (parallel to the rolled margins) are formed by rows of microscopic hairs alternating with reproductive structures. Abundant on the shallow reef flat, this species grows on rock, shells, or coral fragments.

Lobophora variegata   (Lamouroux) Womersley
[fluffy ruffles form]
DICTYOTACEAE, DICTYOTALES

There are several distinct growth forms of this light brown alga that have not been named in the scientific literature to date. We refer to this form as "fluffy ruffles" because a clump of it (up to 15 cm in diameter) **resembles ruffled loose-leaf lettuce** with lobed margins on the broad blades. Characteristically a plant of calm, shallow waters (1–8 m) of low herbivory, often found among seagrasses. [*Pocockiella variegata* in Taylor (1960); see Womersley (1967)]

Lobophora variegata   (Lamouroux) Womersley
[shelf form]

DICTYOTACEAE, DICTYOTALES

This second growth form is represented by thinner, **shelflike,** overlapping, fan-shaped, prostrate blades up to 15 cm in diameter, in some ways resembling flat potato chips. Found in shaded shallow areas or in deep-water habitats of moderate herbivory, this species is often the dominant plant at a depth of 100 m.

Lobophora variegata   (Lamouroux) Womersley
[crust form]

DICTYOTACEAE, DICTYOTALES

The third form is a darker, **orange-brown, encrusting** plant that tightly adheres to dead coral, mangrove roots, or sunken logs in shallow subtidal areas (to 30 m) where grazing is intense.

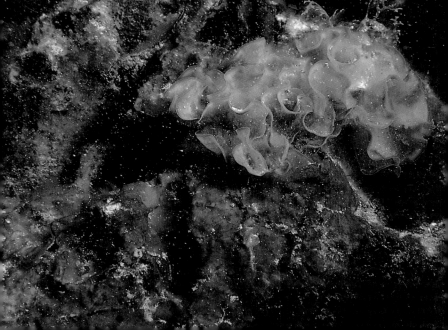

Cladosiphon occidentalis   Kylin
CHORDARIACEAE, CHORDARIALES

These slender, cylindrical, sparsely branched plants are **dull brown in color, soft, gelatinous and slippery** in texture, and covered with fine, colorless filaments or hairs. Up to 30 cm tall and irregularly branched, individuals are attached by a small, inconspicuous, disk-shaped holdfast to other algae or, more commonly, seagrasses in relatively shallow (less than 11 m) quiet waters. [*Eudesme zosterae* in Taylor (1960); see Earle (1969) for further information]

Hydroclathrus clathratus   (C. Agardh) Howe
SCYTOSIPHONACEAE, SCYTOSIPHONALES

Spherical and hollow early in its development, this species later becomes irregularly shaped, characteristically resembling a spongy Swiss cheese, with **many distinct holes of varying sizes** throughout the plant. "Clathrate" means "netlike," and some forms resemble a highly perforated *Colpomenia*. Light to pale golden brown, *Hydroclathrus clathratus* may grow to 1 m or more in length with no apparent holdfast; it typically occurs in relatively cool waters to a depth of 10 m.

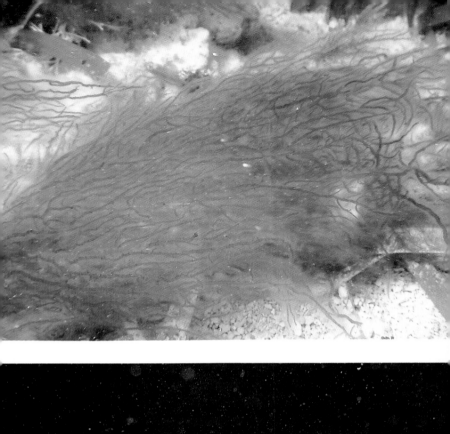

Colpomenia sinuosa   (Roth) Derbes and Solier in Castagne
SCYTOSIPHONACEAE, SCYTOSIPHONALES

Smooth, slick, hollow, golden brown plants that are spherical to irregular in shape and crisp in texture (*Colpomenia* means "sinuous membrane"). Plants may reach 30 cm in diameter and can be 10 cm tall with no single apparent holdfast. Attachment occurs at many points, and individuals may extend into colder northern regions, growing as epiphytes on other organisms or on any solid substrate from the intertidal region to 15 m deep.

Chnoospora minima   (Hering) Papenfuss
CHNOOSPORACEAE, SCYTOSIPHONALES

The distinguishing feature of this alga is the slightly **broadened and flattened area where each branch divides.** Brown in color and up to 15 cm tall, this wiry, tough, thin, dichotomously branched species is attached tightly to rock by a disk-shaped holdfast. *Chnoospora minima* is found in the high-intertidal zone in areas exposed to relatively large waves.

Cystoseira myrica   (Gmelin) C. Agardh
CYSTOSEIRACEAE, FUCALES

Somewhat resembling a sparse, thin *Sargassum*, but lacking leaf-like blades, this species is wiry, stringy, and light brown in color, with irregular to alternate branching. Plants can reach a height of 30 cm and possess **floats or bladders on the outer portions of the branches** (*Cystoseira* means "bladder chain"). Individuals are patchily distributed but can be locally abundant on rocks 1–10 m deep. This species is reported only from the Bahamas and Florida.

Sargassum natans   (Linnaeus) Gaillon
SARGASSACEAE, FUCALES

One of the two species of *Sargassum* characteristic of the Sargasso Sea and often called sargasso weed, it forms free-floating clumps with pale brown, elongated, narrow blades and spiny margins on its thin, wiry branches. Numerous spherical bladders are connected to the main stalk by elongated "stems," and the blades lack cryptostomata (scattered, small, dark dots observed when held up to a bright light). The distinguishing character is the **small spine, hooked spur, or leaflike projection** that appears at the tip of each bladder. Plants generally have no single, distinct, main axis or branch but are composed of clusters of branches. Individuals may reach a length of 50 cm and are often found floating alone or in large clumps or rafts, often becoming major components of beach drift.

Sargassum fluitans   Boergesen
SARGASSACEAE, FUCALES

Sometimes called gulf weed, this species looks more robust than *Sargassum natans,* with round floats or bladders along the smooth stem; individual plants may reach lengths up to 1 m. The golden brown "leaves," though numerous, are not dense on the widely spreading cylindrical branches. These short-stalked, thin, flat blades have a prominent midrib and pointed teeth on the margins; they lack cryptostomata (scattered, small, dark dots that are visible when the blades are held to the light). This species also is strictly free-floating in open ocean waters and is often found in large clumps or rafts on surface waters; usually a major component of beach drift.

Sargassum hystrix   J. Agardh [var. *hystrix*]
SARGASSACEAE, FUCALES

A rather bushy plant, occasionally reaching a height of 40 cm, with a **white midrib** (center vein) apparent on the dark brown, oval, living leaves. Stems are smooth and cylindrical and attached to rocks or corals by a strong holdfast. Uncommon in shallow waters, it is frequently encountered in deep waters (10–100 m). This species is the deepest-growing brown alga known, having been observed at depths of 115 m in the Bahamas.

Sargassum hystrix var. *buxifolium*   Chauvin in
J. Agardh
SARGASSACEAE, FUCALES

Generally considered a variety of *Sargassum hystrix*; some experts think it may be a separate species. It **lacks the distinctive white midrib** of *Sargassum hystrix* but shares many of the other characteristics, such as the large (up to 6 cm long and 1.5 cm wide), dark brown, **crowded, oval "leaves"** with small cryptostomata (tiny dark spots on the blades). The plants are usually taller (up to 50 cm) and more frequently found in shallow waters (1–15 m) attached to rocks or other hard substrates by a sturdy holdfast.

Sargassum polyceratium   Montagne
SARGASSACEAE, FUCALES

A large species up to 1 m long, often resembling a densely branched, brown bush. Tough and leathery with a small but strong holdfast attached to rock, this species has one to several primary stalks with many branches (younger stalks usually are very spiny) and numerous but relatively small (up to 2 cm long) leaflike blades. The basal margin of the blade is occasionally flat or perpendicular where the stem attaches, and the berrylike floats are located near the bases of the blades. Commonly found in shallow (1–5 m) or intertidal habitats just behind the reef crest, plants also occur as deep as 10 m in protected areas. In China, *Sargassum* is an important economic seaweed, producing medicinal drugs and used as animal feed and fertilizer.

Sargassum platycarpum   Montagne
SARGASSACEAE, FUCALES

The golden brown, leaflike blades are either sparsely scattered on the smooth, slender main branches or crowded on short branchlets and have margins or edges with large toothlike projections. Air bladders appear on short stalks, and mature plants may reach a height of 40 cm. The base of the blade gradually tapers into the stem, and **large, easily seen cryptostomata** (dark dots on the blade surface) are arranged in irregular rows parallel to, and on each side of, the midrib (center vein). Plants commonly occur in shallow waters (1–5 m) in areas exposed to moderate wave shock.

Sargassum pteropleuron   Grunow
SARGASSACEAE, FUCALES

One of the largest Caribbean marine plants, this species reaches **heights of 2–4 m,** appearing as vertical columns in the water. Each of the light brown, short-stalked blades (up to 8 cm long) has a toothed margin and a clearly distinguishable raised midrib, appearing as a triangle in cross section with teeth or other projections arising from the midrib ridge; the young blades at the end of the branches also often appear triangular. *Sargassum pteropleuron* contains scattered cryptostomata (dark spots where sterile hairs are produced) on both the blades and stems. The cylindrical stems are alternately branched and buoyed by exceptionally large spherical bladders on short stalks. These impressive plants grow in shallow waters (about 1–5 m deep), usually attached to small rocks or coral fragments by a substantial holdfast.

Turbinaria turbinata   (Linnaeus) Kuntze
CYSTOSEIRACEAE, FUCALES

A handsome, tall (40 cm), erect, cylindrical, brown plant with branches that bear clusters of pyramid-shaped "leaves" (triangular in cross section and tapering toward the stalk). The tough and leathery smooth-edged "leaves" are sometimes sharply spined at the corners, and each contains a **slight swelling (embedded air bladder) in the center of the triangular, flat or concave (slightly sunken) top.** The strong holdfast adheres tightly to rock behind the reef crest in waters ranging from intertidal to 5 m deep in areas of strong turbulence. Species of *Turbinaria* are eaten raw or pickled, as well as used for fertilizer on coconut plantations.

Turbinaria tricostata   Barton
CYSTOSEIRACEAE, FUCALES

Similar to *Turbinaria turbinata,* this species differs in having small sharp teeth on the stalk or tapered portions of the pyramid-shaped blades; also, it **lacks the swollen areas on the smooth, concave tops of the blades.** Brown plants stand erect to 40 cm tall and are sometimes found intermixed with *Turbinaria turbinata* and *Sargassum polyceratium* behind the reef crest in shallow areas with strong currents (intertidal zone to 5 m deep).

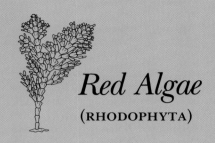

Red Algae

(RHODOPHYTA)

The Rhodophyta (from the Greek *rhodon* for "red rose") is by far the largest and most diversified group of tropical reef plants, with more than 4,000 species. Rhodophyta occupy the entire range of depths inhabitable by photosynthetic plants, from high-intertidal regions to depths as great as 268 meters. Their range of forms is extraordinary, and they can occur in almost any color imaginable, making them the most beautiful and appealing of all algae. The characteristic color is some shade of red, which is the result of large quantities of the pigment phycoerythrin. The precise identification of many red algae depends upon microscopic reproductive features that are difficult to find and interpret. The species depicted herein, however, can be recognized for the most part from external features. Most Caribbean species are not well understood, and many have not yet been named, as exemplified by the mystery alga pictured in the section on Rhodophyta.

Few people appreciate that red algae dominate and often exceed corals in importance as reef-building organisms. In fact, without certain species of calcareous reds, most reefs would not exist. Instead of "coral reefs," we prefer the term "biotic reefs" or just "reefs" when referring to these complex marine structures, since the coral animal's role is mainly that of providing the bulk material. Coralline red algae can form an algal ridge that absorbs wave energy and thereby protects the more delicate corals, fleshy algae, sponges, and other organisms that inhabit the sheltered lagoons and back-reef habitats.

Martensia pavonia   (J. Agardh) J. Agardh
DELESSERIACEAE, CERAMIALES

This alga has wide, thin, delicate, pale, translucent, bluish pink lobes usually emerging from a mat of other algal species. Plants consist of a membrane without veins and having no stalk. The membranous blades (1–3 cm long) are divided into zones; **every other zone is a gridwork of elongated holes in the one-cell-thick, netlike blade,** followed by a zone of solid membrane several cells thick. This species is found from shallow to moderate depths (to 30 m); other species in this genus have been found as deep as 100 m.

Halymenia floresia   (Clemente) C. Agardh
GRATELOUPIACEAE, CRYPTONEMIALES

Showy, bright, pinkish, translucent red, large plants that grow to 50 cm tall with thin, wide blades (main blades up to 3 cm wide). Branching is irregular to pinnate with numerous, small, narrow branchlets on the margins. Attached to hard substrates by a small holdfast, this species is characteristically a deep-water form (5–40 m).

Halymenia floridana   J. Agardh
GRATELOUPIACEAE, CRYPTONEMIALES

This attractive plant, ranging in color from red to yellow, grows up to 20 cm long and has a gelatinous, slippery, fleshy consistency to the smooth, thick blades. Blade margins are lacey or undulating. An uncommon species, usually attached to hard substrates with a small holdfast and small, inconspicuous stipe in 5–15 m of water.

Halymenia duchassaignii   (J. Agardh) Kylin
GRATELOUPIACEAE, CRYPTONEMIALES

A large (20 cm tall), striking plant, highly variable in color from mottled red to cream. Blades are fleshy and wide, with irregular lobes that have upward-pointing teeth on the undulating margins. The **blades possess many conspicuous, dark, rounded bumps or projections,** giving the surface a mottled, rough, rubbery appearance, while still having the slippery feel of a highly gelatinous blade. Found as scattered individual plants on reef flats 1–7 m deep.

Kallymenia limminghii   Montagne
KALLYMENIACEAE, CRYPTONEMIALES

Small (up to 5 cm) bright red plant with strap-shaped blades having rippled or undulating margins; occasionally branched but more often not. This species is found in low-light habitats such as caves and cracks, beneath rock ledges, or, typically, at the bases of fan and whip corals in waters from the low-tide mark to 30 m.

Anotrichum barbatum   (J. E. Smith) Nageli
CERAMIACEAE, CERAMIALES

Plants are composed of small, erect tufts of filaments, usually 1–5 cm (rarely to 10 cm) tall, that are bright rose in color. When examined closely, each filament consists of a **chain of small, swollen, elongated cells** that taper slightly at the ends. When present (as pictured), reproductive structures appear as dark, round bodies within specialized cells. Tufts may occur on stones or other hard objects in shallow waters (above 5 m). The genus *Griffithsia*, a close relative, differs in microscopic reproductive characteristics and has larger, more oval, oblong cells. [*Griffithsia barbata* in Taylor (1960); see Baldock (1976) for further information]

Griffithsia globulifera   Kuetzing
CERAMIACEAE, CERAMIALES

Forming a delicate, easily crushed component of mixed algal turfs, these small, bright rose, almost transparent plants stand erect to 6 cm tall. All parts of the plant body appear similar, with **no distinction between main branches and branchlets.** The uprights consist of strings of beadlike cells held firmly to rocks by a small fibrous holdfast, or they grow epiphytically on larger, coarse algae in shallow waters (1–10 m).

Callithamnion cordatum   Boergesen
CERAMIACEAE, CERAMIALES

Callithamnion means "beautiful branch," and individual plants are small (2–4 cm tall), bushy, and pale bluish pink. The branches generally divide alternately and possess many small branchlets that appear as fine, sparse hairs. Individuals are epiphytic and attach to other algae by small, inconspicuous rhizoids; present from shallow waters to a depth of 30 m.

Champia parvula   (C. Agardh) Harvey
CHAMPIACEAE, RHODYMENIALES

Champia parvula can grow to 10 cm tall and is characterized by yellowish to pale pink, narrow, cylindrical, alternate branches that are partitioned into large, hollow segments by thin, membranous diaphragms (this character readily distinguishes the genus). The branches are more slender **(normally less than 1.5 mm in diameter)** and are less swollen than those of *Champia salicornioides,* with tips that taper to blunt points. Small individuals frequently occur as inconspicuous epiphytes on seagrasses in relatively shallow waters (1–15 m deep).

Champia salicornioides   Harvey
CHAMPIACEAE, RHODYMENIALES

A small (up to 12 cm tall) but robust pale red plant composed of thick, hollow, gelatinous, irregular branches partitioned into short, barrel-shaped, swollen segments by thin diaphragms. Each segment is about as wide as it is long **(2–4 mm in diameter).** This alga is occasionally found attached to hard substrates by a small, inconspicuous holdfast in moderately shallow waters 1–10 m deep.

Centroceras clavulatum   (C. Agardh in Kunth) Montagne
in Durieu de Masionneuve
CERAMIACEAE, CERAMIALES

Dark brownish maroon plants that typically grow as filamentous mats or tufts up to 20 cm long. Branches are dichotomous and, upon close examination, contain **alternating light and dark bands that result from minute spiny projections** (easily seen with a hand lens) separating the lighter segments (*Centroceras* literally means "horns around a center"). The branch tips are incurved, looking like pincers or lobster claws. Mats or tufts of this alga may be found on hard substrates from the intertidal zone to 5 m deep throughout the Caribbean.

Ceramium sp.
CERAMIACEAE, CERAMIALES

There are many species within this genus, most of which are recognized by distinct **banding patterns of small cells (no spines) on the fine cylindrical filaments** (usually just visible to the unaided eye) and also by the characteristic incurved branch tips. However, it is practically impossible to identify most *Ceramium* to the species level with the unaided eye, and microscopic examination is often insufficient; one has to locate the reproductive structures. Therefore, we have chosen to depict only the generic characteristic of banding. This widespread and often common genus can be found on rocks, in tide pools, and as epiphytes on other plants or on most firm substrates.

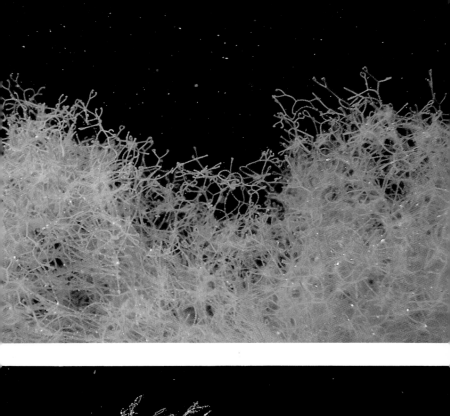
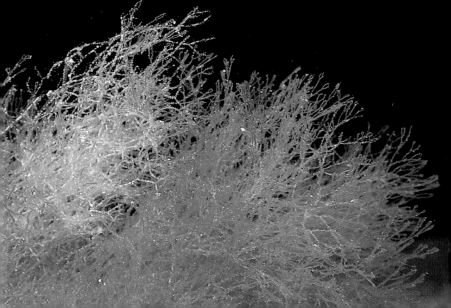

Ceramium nitens   (C. Agardh) J. Agardh
CERAMIACEAE, CERAMIALES

A few members of this genus have a growth form distinctive enough to be identified by the unaided eye. *Ceramium nitens* is one such species and forms bright rust- or rose-colored tufts (up to 10 cm tall) with dichotomous branching. The filaments are completely covered with pigmented cells, obscuring the characteristic banding in this species. Plants are common in shallow waters (1–10 m) on dead corals or as epiphytes on other algae.

Wrangelia argus   (Montagne) Montagne
CERAMIACEAE, CERAMIALES

Small (1.0–1.5 cm tall), soft, **iridescent, purple-red, turf alga with whitish gray tips** on the branches. Frequently covering large areas, this species often forms fuzzy, inconspicuous clumps or turfs on rocky substrates in shallow (1–10 m) turbulent areas.

Wrangelia penicillata   (C. Agardh) C. Agardh
CERAMIACEAE, CERAMIALES

Plants are much larger than *Wrangelia argus,* growing to 20 cm tall and developing a distinctive, light pink, bushy form. Branching is alternate in two ranks or in one plane (branches protruding from opposite sides of the main stalk). Individuals are most often found on large, coarse plants but can be attached to any hard substrate in shallow to moderately deep waters (1–15 m).

Heterosiphonia gibbesii   (Harvey) Falkenberg
DASYACEAE, CERAMIALES

Sparsely branched, but each plant looks like a small, delicate, soft, pink, puffy cloud because of the abundance of minute dichotomous branchlets at the ends of each branch. The color is bright pinkish red, and plants grow to 20 cm tall in sunny, wave-protected, shallow waters (less than 6 m deep) on hard substrates.

Spyridia filamentosa   (Wulfen) Harvey in Hooker
CERAMIACEAE, CERAMIALES

A dull, pale pink, bushy plant 15–20 cm tall that arises from a small holdfast. Main branching is alternate with many delicate branchlets, almost filamentous in growth form, which gives a fuzzy appearance to the outer branches. *Spyridia filamentosa* occurs in warm, quiet, protected areas down to depths of 8 m.

Spyridia hypnoides   (Bory in Belanger) Papenfuss
CERAMIACEAE, CERAMIALES

Rose red filamentous plant that can become moderately large, up to 25 cm tall. Alternately branched in any and all directions; upon close examination, the branches seem to be banded with abundant, fine, radially arranged branchlets. Occasionally a relatively large flattened hook is present on the tips of some of the branches. This species is highly variable, depending on the habitat; plants in areas exposed to wave action are generally smaller and more compact, and those in calmer waters are much longer. Present from the low-tide mark to 5 m deep on rock. [*Spyridia aculeata* in Taylor (1960); see Papenfuss (1968) for update]

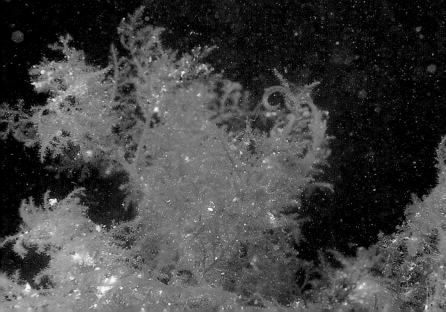

Asparagopsis taxiformis   (Delile) Trevisan
BONNEMAISONIACEAE, BONNEMAISONIALES

A light, pale red to reddish gray delicate plant to 20 cm tall, with the upright main branches composed of such finely subdivided branchlets that the frond takes on a fluffy appearance and soft texture. The main axes arise from horizontal rhizomes, with fibrous holdfasts often forming dense patches at shallow to moderate depths (2–15 m) where there is considerable water motion or current. *Asparagopsis taxiformis* is a favorite *limu* (seaweed) of Hawaiians and is traditionally used to flavor foods such as fish and meat dishes.

Dasya baillouviana   (S. G. Gmelin) Montagne
DASYACEAE, CERAMIALES

A beautiful alga, up to 50 cm tall, that is delicate, bright red, and densely surrounded by fine, hairlike branchlets covering the long, graceful branches. This plant, often called chenille seaweed, is usually alternately branched and slippery to slimy in texture, with a disklike holdfast that attaches to hard substrates in shallow (1–10 m) protected areas. [*Dasya pedicellata* in Taylor (1960); for a detailed account of its name change see Dixon and Irvine (1970)]

Dasya harveyi   Ashmead in Harvey
DASYACEAE, CERAMIALES

Plants are rose red, up to 25 cm tall, and bushy with no dominant, main stalk but many primary alternate branches. The small branchlets are covered with randomly arranged hairs or fine filaments. Individuals may be found growing on hard substrates down to 10 m deep.

Eupogodon antillarum   (Howe) Silva in Silva, Menez, and Moe
DASYACEAE, CERAMIALES

An attractive pale brownish purple species up to 12 cm in length. Main branches are purple and irregularly branched, and the outer branches are lighter, alternately divided, and covered with fine, hairlike branchlets. Plants are found attached to solid substrates in shallow waters (less than 10 m deep). [*Dasyopsis antillarum* in Taylor (1960); see Silva, Menez, and Moe (1987) for the name change]

Chondria tenuissima   (Goodenough and Woodward)
C. Agardh
RHODOMELACEAE, CERAMIALES

Small (up to 20 cm), inconspicuous, pale straw-colored to dull purple plants. The main axes are relatively coarse with few branches, and the outer parts are somewhat finer with more branches. Branchlets are usually simple (unbranched) and are **tapered at both the attached end and the outer tip.** When viewed with a hand lens, the outer tip shows a tuft of fine filament. Plants occur in the intertidal or very shallow subtidal zones (less than 1 m), growing on boat ramps, stones, or shells.

Chondria littoralis   Harvey
RHODOMELACEAE, CERAMIALES

A pale to deep red species, reaching 30 cm tall. Branching is irregular, with the main branches 1–2 mm in diameter. Branchlets are much smaller and distinctly **pinched at the base. The blunt tips reveal a tuft of very fine filaments** when closely examined with a hand lens. This plant is most commonly encountered 2–15 m deep and intermingled with other small algae as thick, short (1–5 cm) turfs.

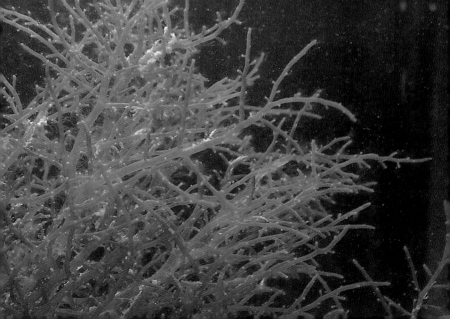

Dudresnaya crassa   Howe
DUMONTIACEAE, CRYPTONEMIALES

Erect, bushy, highly gelatinous, **gooey, translucent, gray-red plants,** 5–20 cm tall. A small cylindrical stalk (to 4 mm in diameter) gives rise to many densely branched fronds. This slimy, fragile alga is found 1–20 m deep on coral fragments and is generally uncommon, but when present (usually appearing in the springtime), it may occur in considerable abundance.

Scinaia complanata   (Collins) Cotton
CHAETANGIACEAE, NEMALIALES

Plants are quite distinctive, **translucent, pale red, and usually 5–8 cm tall. Branching is strictly dichotomous.** Regularly cylindrical branches with a soft gelatinous texture are all approximately the same length, creating a bushy, hemispherical plant. This alga grows at moderate depths (2–10 m) on coral fragments.

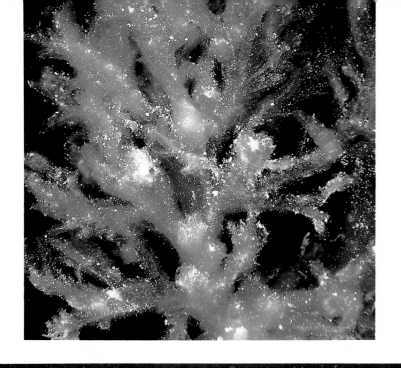

Trichogloeopsis pedicellata   (Howe) Abbott and Doty
HELMINTHOCLADIACEAE, NEMALIALES

This stark white plant (due to moderate calcification) is soft, fleshy, and flaccid and can grow to 16 cm tall. Often confused with *Liagora;* however, it is much more gelatinous and has alternate branching. Found on rocks or coral fragments, this alga occurs at shallow to moderate depths (1–12 m) and is often encountered in the groove portion of the spur-and-groove area seaward of the reef crest. [*Liagora pedicellata* in Taylor (1960); see Abbott and Doty (1960) for name change]

Liagora mucosa   Howe
HELMINTHOCLADIACEAE, NEMALIALES

Soft, gelatinous (gooey) plant reaching a height of 20 cm. The pinkish white color is the result of light calcification. Individuals grow in loose, irregularly branched clusters in moderately shallow waters (to 15 m deep) on rocks or coral rubble.

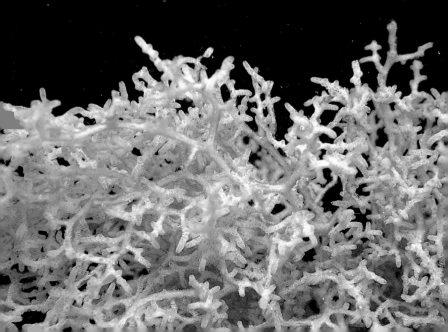

Liagora ceranoides   Lamouroux
HELMINTHOCLADIACEAE, NEMALIALES

Moderate calcification imparts a white color to these compact plants that can grow to 15 cm tall. The sparsely branched stalk gives rise to arched main branches; the **outer portions have numerous short, widely angled, dichotomous branchlets,** creating a soft, hemispherical clump. The lower portions of the main stalks appear bumpy or scaly. Individuals may be found on rocks or coral fragments to 20 m deep.

Liagora farinosa   Lamouroux
HELMINTHOCLADIACEAE, NEMALIALES

Rather small, soft, delicate plants with a whitish red color (whiteness caused by light calcification) that grow to a height of 10 cm. Branching is somewhat dichotomous. Populations occur in shallow waters on rock or coral fragments usually in sandy sheltered areas to a depth of 10 m.

Liagora pinnata   Harvey
HELMINTHOCLADIACEAE, NEMALIALES

Plants are soft to the touch, small (10 cm tall), and pinkish, with irregular opposite branches that form a moderately dense unit. Individuals tend to be lightly calcified and are commonly found on rocks or coral fragments in shallow calm waters (1–10 m), often growing in the same habitat with *Liagora farinosa*.

Trichogloea herveyi   W. Taylor
HELMINTHOCLADIACEAE, NEMALIALES

Gelatinous, slippery, slimy, cylindrical, purplish red plant (to 30 cm tall) that is soft and gooey out of water but strikingly attractive underwater. Unlike *Trichogloea requienii*, there is little or **no visible calcification** on its bushy, alternate, blunt branches, with main axes up to 8 mm in diameter. Found from the low-tide mark to 10 m deep, individuals are attached to rocks or other hard substrates; uncommon, but abundant when present.

Trichogloea requienii   (Montagne) Kuetzing
HELMINTHOCLADIACEAE, NEMALIALES

Slimy, gooey species; the branches are white from light calcium carbonate deposition (lime deposits). The outer portion is surrounded by a bright red, gelatinous halo that creates a distinctively attractive and gracefully flowing effect. Individuals grow to 20 cm tall and are bushy with alternate branching; **calcified main branches are as much as 4 mm in diameter.** This alga is attached to hard substrates by a small holdfast in shallow or moderately shallow (1–15 m) waters.

Unknown red alga

This plant is quite an oddity. Soft, gooey, highly gelatinous, and extremely slippery (slips through your fingers when held out of water), this light pink alga is a shapeless blob in air, but when submerged, the branches extend randomly from the thick, somewhat flattened main stalks. Growing to a height of 8 cm, individuals can be found between the branches of corals 2–6 m deep. The presence of calcium carbonate within this plant makes it particularly unusual among similar gooey genera.

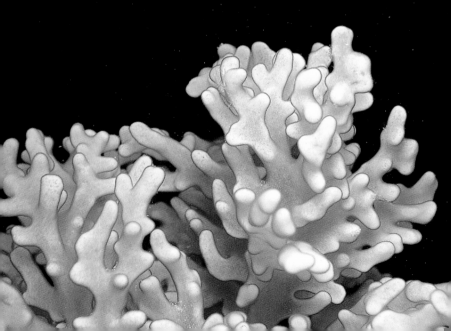

Dictyurus occidentalis   J. Agardh
DASYACEAE, CERAMIALES

A small (up to 12 cm tall), erect, dark red seaweed. Sparingly branched, but most of each branch is covered by short, radially arranged (whorled) branchlets, which give rise to even **finer branchlets that form a lacy web of quadrangles** (a distinctive feature of this genus); the net of branchlets creates a rather soft, delicate, four-sided plant. Found infrequently at shallow to moderate depths (1–10 m), where it can be locally abundant.

Haloplegma duperreyi   Montagne
CERAMIACEAE, CERAMIALES

Composed of pale pink, spongy, ruffled, lobed fronds, this plant characteristically lies flat on the substrate. Patches grow to 12 cm wide but only 5 cm tall. **Densely interwoven filaments (without a main axis or vein)** form the flattened irregular blades and make the patches look like felt. Plants are generally found in the shade under overhanging rock ledges or at the bases of larger algae at depths of 1–15 m.

Coelothrix irregularis   (Harvey) Boergesen
CHAMPIACEAE, RHODYMENIALES

Plants often grow as a conspicuous **entangled, wiry, bright iridescent blue turf,** about 2–3 cm tall, and form sparse to dense mats of irregularly branched, slender, rigid uprights. These mats, or turfs, can be found from the rocky intertidal zone to 10 m deep in shaded areas of cracks and crevices or under ledges.

Gelidium pusillum   (Stackhouse) Le Jolis
GELIDIACEAE, GELIDIALES

Small (2–5 cm), wiry, tough, dull dark purple, turf-forming plant with primary branches arising from rhizomes that are firmly attached to rocky substrates. Uprights are strap-shaped, often with pinnate branchlets. Commonly found from the intertidal region to 8 m deep, and because of its tough wiry nature, this alga can withstand considerable wave shock. *Gelidium* means "gelatin" (from *gelu*); some species are commercially valuable for the production of agar, used in many food and pharmaceutical industries and specifically for culture media in the medical sciences. [Dixon and Irvine (1977) combined *Gelidium crinale* and *Gelidium pusillum,* which were treated separately by Taylor (1960)]

Gelidiella acerosa   (Forsskal) J. Feldmann and Hamel
GELIDIACEAE, GELIDIALES

Individuals are rhizomatous, wiry, and quite tough, with slender opposite branches. They grow to 15 cm tall. Color is greenish yellow to dark brown. Tips of the sparse branches are often recurved or end in an arch. Found in areas of heavy fish grazing, where the plant intertwines with other species to form sparse turfs (1–3 cm high) covering various hard substrates; often the main component of mixed algal turfs. Plants are found from intertidal habitats to 7 m deep on rock or dead coral.

Bostrychia montagnei   Harvey
RHODOMELACEAE, CERAMIALES

Blackish or dull dark purple when wet, pale yellow on upper surfaces when dry. Occasionally plants may be found as tall as 8 cm but are normally smaller. This species has distinctive, bilateral (offshoots on both sides), featherlike branches with curled tips. It most commonly forms **tufts that droop or hang loosely when exposed at low tide.** Lower portions of the main branches have few if any branchlets. Plants grow almost exclusively in the upper intertidal zone on mangrove prop roots in sheltered situations, but are occasionally found on rocks, pier pilings, and seawalls.

Bostrychia tenella   (Vahl) J. Agardh
RHODOMELACEAE, CERAMIALES

Plants are small (2–5 cm tall) and most often found as nearly black, dense, mosslike clumps, although dark red to dark purple individuals may have runners covering distances of 5–10 cm. When separated and examined closely, plants show sparse branching near the tangled rhizoidal base as well as dense bilateral branching above, toward the typically inwardly curled tips. Individual clumps or mats most frequently occur on intertidal rocks, seawalls, or other hard substrates as well as on mangroves, usually in shaded localities.

Wrightiella blodgettii   (Harvey) Schmitz
RHODOMELACEAE, CERAMIALES

Attractive, stiff, spiny, bright red plant (to 20 cm tall) composed of numerous branches, each with **four rows (ranks) of small spurlike branchlets** 1–5 mm long. On some plants, the spurs occasionally may be forked. This alga can be found attached to hard substrates by an inconspicuous holdfast and has been reported growing to 36 m deep.

Ochtodes secundiramea   (Montagne) Howe
RHIZOPHYLLIDACEAE, CRYPTONEMIALES

Small (4–7 cm tall), bushy, **iridescent, bright blue or purple** alga with numerous alternate branches. Each individual branch grows in one plane or appears flat; however, the whole plant, consisting of many branches, forms a bushy clump. This species is found in moderately turbulent areas 1–15 m deep.

Hypnea cervicornis   J. Agardh
HYPNEACEAE, GIGARTINALES

Hypnea cervicornis has a wiry, entangled growth form and is bright yellow in areas of strong sunlight but darker brownish red in shaded habitats. The size varies greatly, ranging from 3 cm to 30 cm tall. The main branches have many short, pointed, tendril-like branchlets. Plants attach by holdfasts to rocks or coral substrates but can also grow as epiphytes on other larger seaweeds down to 10 m deep.

Hypnea musciformis   (Wulfen in Jacquin) Lamouroux
HYPNEACEAE, GIGARTINALES

The most characteristic feature of this alga is that many of the **branch tips end in flattened hooks,** with small branchlets arising from their outer curves. The plants are orange-red, have wiry slender branches, and grow to a height of 30 cm. Entangled on other algae or attached by a holdfast to hard substrates, this species is found in shallow sheltered areas to a depth of 12 m.

Digenia simplex   (Wulfen) C. Agardh
RHODOMELACEAE, CERAMIALES

The main erect branches are small (2–25 cm tall) and covered with numerous **short (3–5 mm), stiff, wiry branchlets** that are light pink to dull, dark, brownish red. Found from the lower intertidal zone to 20 m deep, they attach to hard substrates, often in heavy surf conditions but sometimes partly buried by sand. Specimens can be clean, as in this photograph, or, more often, overgrown by filamentous epiphytic species that obscure the plant's brushlike aspect.

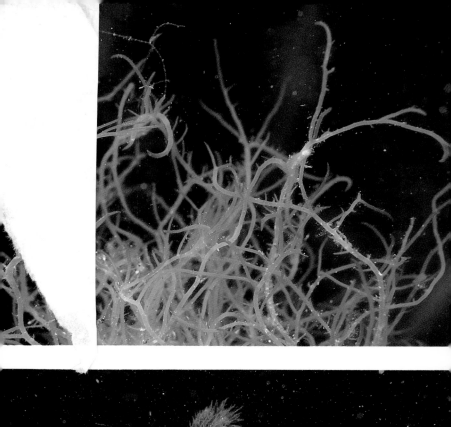

Laurencia intricata   Lamouroux
RHODOMELACEAE, CERAMIALES

Erect plant with a long, sparsely branched, cylindrical, greenish yellow, main stalk containing **rose-colored stubby branchlets.** Tall (15–25 cm), sparse individuals have alternate branching on the main axis, but the tips show opposite branching. Branches occasionally occur in ranks of four when viewed from the tip toward the base. Commonly found in sheltered sandy areas, this species grows on rocks, shells, or coral fragments about 1–3 m deep. The fleshy genus *Laurencia* is named for the French naturalist De la Laurencie and is characterized by an indentation or pit on the tip of each blunt branchlet.

Laurencia poitei   (Lamouroux) Howe
Rhodomelaceae, Ceramiales

Pale buff to **bright pinkish red,** this alga can reach 10 cm tall. Branching is alternate, arising from a stiff main axis with some branches partially flattened. The tip branchlets are short and wartlike and may be opposite or alternate. This species reaches peak abundances in wave-surge areas on rock substrates in the spur-and-groove area of the fore reef 1–17 m deep.

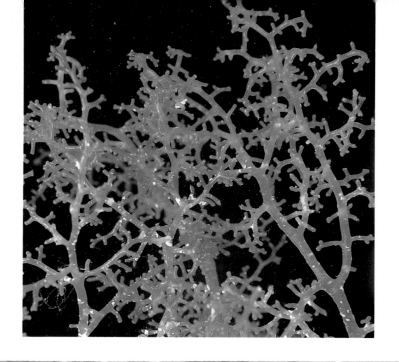

Laurencia obtusa   (Hudson) Lamouroux
RHODOMELACEAE, CERAMIALES

Plants are of medium size (10–20 cm), a peculiar **dark bright green** in color, and somewhat compact or clumped in growth form. Tips are alternately branched. Common in shallow wave-washed areas and strong currents, this species occurs to depths of 8 m. *Laurencia obtusa* contains several strong chemicals that deter herbivorous fishes and sea urchins.

Laurencia papillosa   (C. Agardh) Greville
RHODOMELACEAE, CERAMIALES

This moderately tall (up to 16 cm) green or greenish purple plant forms dense clusters. Branching may be either alternate or irregular. **Tips are densely covered with short, tough, knobby branchlets;** the lower parts of branches may be devoid of branchlets. Growing on hard substrates exposed to moderate waves from the intertidal zone to 7 m deep, this species is common and widespread throughout the Caribbean.

Mystery alga
FAMILY UNKNOWN; ORDER UNKNOWN

We refer to this plant as the "mystery alga" because it is so peculiar that no scientific name has been assigned to it, no genus has been designated, and even the order to which it belongs is in question. Tough and leathery with a slick feel to the fronds, the plants are pinkish white to brownish red and grow to 30 cm tall. It is occasionally found in fore-reef habitats but has been overlooked in the past because of its close resemblance to gorgonian corals (common sessile animals on reefs). Generally found on dead coral or rock, this alga occurs 2–20 m deep.

Flahaultia tegetiformis   W. Taylor
SOLIERIACEAE, GIGARTINALES

This small (3–5 cm in diameter) but attractive **bright maroon plant is fairly tough and rubbery, with lobed blades; the plant body spreads close to the substrate** and is attached by a small holdfast connected to the center of the blade. Because it occurs in dark caves and crevices from shallow to fairly deep waters (1–30 m), this inconspicuous alga is seldom found. [see Taylor (1974) for the technical treatment]

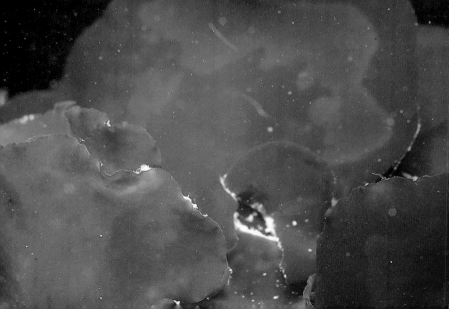

Chrysymenia sp.
RHODYMENIACEAE, CRYPTONEMIALES

This completely hollow species can grow to 30 cm tall. Its wide cylindrical to slightly compressed light rose red fronds can show opposite, alternate, irregular, or even dichotomous branching. The branches are pinched or narrow where they attach to the main upright. This moderately deep-water plant (15–40 m) attaches to hard substrates by a small, inconspicuous holdfast.

Botryocladia pyriformis   (Boergesen) Kylin
RHODYMENIACEAE, RHODYMENIALES

These small (1–2 cm tall) plants grow as wine red, balloon- or grapelike, tear-shaped spheres clustered atop short stalks (generally having **1 or 2 bladders per stalk**) and form small groups of 4–6 uprights that arise from a single holdfast. Usually present in shaded rock cracks and crevices on shallow reef flats; individuals are occasionally found in deeper waters (to 40 m). Another species that may be encountered in the Caribbean, *Botryocladia occidentalis* (Boergesen) Kylin, has similar but smaller bladders that occur abundantly along the longer (up to 25 cm) branches.

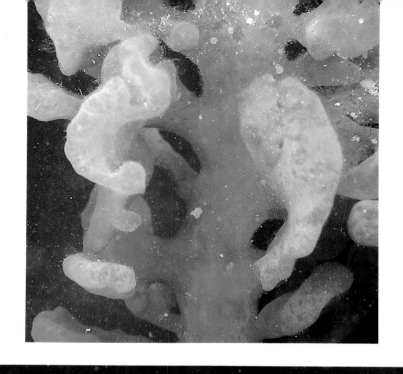

Acanthophora spicifera   (Vahl) Boergesen
RHODOMELACEAE, CERAMIALES

This erect, brittle (fragments easily), irregularly and sparsely branched alga grows to 25 cm tall. Highly variable in color, it ranges from whitish pink to pale brown, green, or even almost yellow. Fine **spinelike branchlets of hairs appear at the tips of short, stumpy, spurlike branches** that uniformly cover the main uprights. This alga is commonly found attached to other organisms, coral fragments, or pebbles in calm waters from the intertidal zone to 8 m deep.

Bryothamnion triquetrum   (S. G. Gmelin) Howe
RHODOMELACEAE, CERAMIALES

Growing to 25 cm tall, the dark brownish red branches appear bushy or shrublike. The species has numerous, coarse, stalklike, alternately branched axes bearing many tough, stiff, **short branchlets that are forked at the tips and arranged in ranks of three.** In end view, the branches are triangular (*triquetrum* means "three-cornered"), often spiraling or twisting. These bushy plants are found on rocks exposed to moderate wave surge in shallow to medium depths (4–15 m).

Gracilaria tikvahiae   McLachlan
GRACILARIACEAE, GIGARTINALES

A highly variable species, ranging from deep green to bright red and up to 30 cm tall. *Gracilaria* means "slender" or "delicate," and some plants form loose (sparse) clumps, but others are much tighter and more branched. The outer branches are cylindrical and slightly flattened. Common in calm estuaries and bays, these plants can grow free or attached to small rocks and coral fragments to 10 m deep. [*Gracilaria foliifera* var. *angustissima* in Taylor (1960); see McLachlan (1979) for update]

Gracilaria damaecornis   J. Agardh
GRACILARIACEAE, GIGARTINALES

A bushy, tough, leathery, pinkish red plant that can grow to 15 cm tall. Distinctive **dichotomous branching** generally is present throughout the plant. **Branches are cylindrical,** not pinched or constricted at the point of attachment, with short blunt tips. This species usually is found in shallow waters (1–10 m) on coral fragments, rocks, or other hard substrates.

Gracilaria domingensis   Kuetzing
GRACILARIACEAE, GIGARTINALES

This rose red plant is generally less than 10 cm tall but may reach 35 cm. The main branches are flattened and **strap-shaped with opposite branchlets emerging from the edges.** The cystocarps (reproductive structures), when present, are obvious as numerous dark red bumps on the flattened branches. Individuals are found on coral fragments and other hard substrates in shallow waters (2–15 m deep).

Gracilaria curtissiae   J. Agardh
GRACILARIACEAE, GIGARTINALES

Gracilaria is a highly variable genus that makes proper identification difficult; this particular species, however, is pale pink to straw-colored and tall (40 cm), with dichotomous branches arising from a broadly flattened basal blade. The **upper, strap-shaped, smooth-edged blades are often constricted (narrower) at the base.** Male reproductive structures appear as light patches on the blades, and female reproductive structures as small, dark red, hemispherical bumps scattered irregularly over the older areas of the frond. Plants occur 4–10 m deep in moderately wave-exposed to protected areas.

Gracilaria mammillaris   (Montagne) Howe
GRACILARIACEAE, GIGARTINALES

This dark red, tough, fleshy alga usually grows in thick patches (to 10 cm tall) with somewhat dichotomous branching. It attaches to rock substrates in shallow waters (less than 18 m), often exposed to moderately heavy surf.

Gracilaria cervicornis   (Turner) J. Agardh
GRACILARIACEAE, GIGARTINALES

This fleshy, slippery, reddish brown plant reaches 25 cm in height. The main branching is irregular and flattened; the branchlets are usually opposite. This species is often found as small, inconspicuous plants in the shallow subtidal region (less than 10 m deep) on small pebbles, shell fragments, or other small objects in areas of moderate wave action. Species of *Gracilaria* ("ogo" is the common name in Japan) traditionally are used in many parts of the world as food. In the United States, a cooler-water species, *Gracilaria verrucosa* (Hudson) Papenfuss, is a source of high-quality agar, the gel on which medical bacterial cultures are maintained.

Polycavernosa crassissima   (P. and H. Crouan in Schramm and Maze) Fredericq and J. Norris
GRACILARIACEAE, GIGARTINALES

Large (30 cm across), rubbery, tough, reddish brown plant with a metallic, gold-silver sheen. Branches are somewhat flattened and broad, often tangled, fused together and curved downward at the tips. The growth form is **not upright but creeping.** Often this alga seems to be clinging tenaciously to the rock substrate. It is most often found on the leeward side of the reef crest. Individuals grow in shallow waters (to 9 m), and the harsher the environmental conditions (high wave shock and air exposure at low tides), the tighter this plant adheres. [*Gracilaria crassissima* in Taylor (1960); see Fredericq and Norris (1985) for an update]

Polycavernosa debilis   (Forsskal) Fredericq and J. Norris
GRACILARIACEAE, GIGARTINALES

Coarse, rubbery, and often gnarled, some individuals are bushy and others sparsely branched (to 25 cm tall). Color is pale straw yellow to pinkish or pale green. The abundant, irregular, somewhat flattened branching varies in size from short and stubby to long and thin. This alga grows on rubble fragments well behind the reef crest in shallow (1–10 m) protected reef-flat areas. In some countries, this plant is rinsed, dried in the sun, broken, and ground, then used to make a porridge or drink that is highly regarded as an aphrodisiac. [*Gracilaria debilis* in Taylor (1960) and *Gracilaria wrightii* in Wynne (1986); see Fredericq and Norris (1985) for further information]

Meristiella echinocarpum   (Areschoug) Cheney and Gabrielson in Gabrielson and Cheney
SOLIERIACEAE, GIGARTINALES

This greenish yellow species generally grows as a low, tangled mass covering an area up to 10 cm in diameter. The strap-shaped, tough, leathery, **flat blades have coarse opposite teeth** or branchlets appearing from the margins or edges. Generally clumps are found creeping low along the substrate in shallow waters 1–10 m deep. [*Eucheuma echinocarpum* in Taylor (1960), see Gabrielson and Cheney (1987) for the revision]

Meristiella acanthocladum   (Harvey) Cheney and Gabrielson in Gabrielson and Cheney
SOLIERIACEAE, GIGARTINALES

A seaweed that forms large clumps up to 1 m in diameter, *Meristiella acanthocladum* is bushy, often entangled, pale yellow to reddish brown, and irregularly branched (from alternate to dichotomous). The texture is firm but fleshy, and the main axes and branches (often more than 1 cm in diameter) are usually slightly flattened with numerous spines or spurs. Uncommon, but when encountered, plants are usually numerous; found in shallow waters 1–40 m deep attached to rocks in areas with a moderate current. According to recent studies, based primarily on reproductive structures, *Meristiella acanthocladum* and *Meristiella gelidium* appear to be the same species; however, since their external features are so distinctive, we list both forms separately. [*Eucheuma acanthocladum* in Taylor (1960), see Gabrielson and Cheney (1987) for the revision]

Meristiella gelidium   (J. Agardh) Cheney and Gabrielson in Gabrielson and Cheney
SOLIERIACEAE, GIGARTINALES

Meristiella gelidium is a tough, thick, and leathery alga up to 10 cm tall; a creamy yellow-rose color when growing in bright light, to pinkish red when found in the shade. Fronds are often flattened, with undulating margins covered by tough, spinelike branchlets that are sometimes forked. Typically uncommon, it grows 1–30 m deep as scattered individuals. According to recent laboratory studies (Cheney and Babbel, 1978; Cheney and Dawes, 1980), *Meristiella acanthocladum* and *Meristiella gelidium* appear to be the same species. [*Eucheuma gelidium* in Taylor (1960), see Gabrielson and Cheney (1987) for the revision]

Eucheuma isiforme   (C. Agardh) J. Agardh
SOLIERIACEAE, GIGARTINALES

Large (30–70 cm tall), conspicuous plants, with color varying from pale straw yellow to light red brown. Branching is highly irregular, with the branches sometimes smooth or often projecting scattered spines; they are always quite tough and snap off with difficulty. This species is found in sheltered shallow areas (to 10 m deep), sometimes loosely entangled among *Thalassia testudinum* (turtle grass). In some countries, this plant is used (as is *Polycavernosa debilis*) to make a porridge or drink that is regarded as an aphrodisiac. Members of *Eucheuma* are among the world's most commercially valuable tropical seaweeds because their colloidal extracts are used to suspend particles and smooth the texture of many dairy products and other foods.

Galaxaura marginata   (Ellis and Solander) Lamouroux
CHAETANGIACEAE, NEMALIALES

A small, mounded seaweed (5–14 cm tall) of loosely compressed blades that are pinkish red and chalky with light calcification. The dichotomous branches often show faint **cross banding near the tips and are flat, with smooth or slightly thickened edges.** Individuals are attached by a single holdfast to hard substrates in tide pools or shallow (1–10 m) reef areas.

Galaxaura oblongata   (Ellis and Solander) Lamouroux
CHAETANGIACEAE, NEMALIALES

These creamy red plants are bushy and about 10–15 cm high, having slender (1–2 mm in diameter), well-calcified (hard), **cylindrical, smooth branches** with flexible joints, becoming very brittle when dried. The numerous dichotomous branches create a compact dome-shaped plant. Found in protected sandy areas attached by a single holdfast to coral fragments or rocks 1–12 m deep. [*Galaxaura cylindrica* in Taylor (1960); see Magruder (1984) for further information]

Galaxaura subverticillata   Kjellman
CHAETANGIACEAE, NEMALIALES

A moderately calcified species with cylindrical, dark red branches that are ringed by minute, hairlike **filaments of two types (long and short) arranged in alternating zones**, creating fuzzy bands. Branching is irregularly dichotomous, often forming hemispherical mounds 4–7 cm high from one point of attachment. Individuals are commonly found in shallow waters 1–10 m deep on coral fragments or rocks.

Jania adherens   Lamouroux
CORALLINACEAE, CRYPTONEMIALES

Jania, named for one of the mythical Greek sea nymphs, is composed of cylindrical, hard, pink, calcified segments connected by flexible joints. It forms small tangled clumps up to 4 cm tall. Each segment of this species is generally 2–4 times as long as wide. Branching is **dichotomous with wide angles (greater than 50°)**. This alga can be encountered growing on hard substrates or other marine plants to a depth of 18 m.

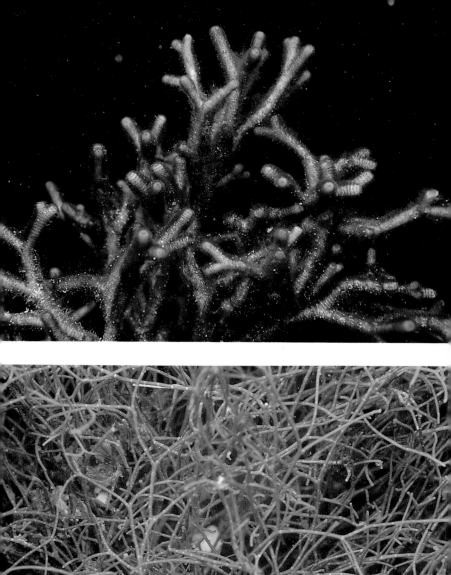

Jania rubens   (Linnaeus) Lamouroux
CORALLINACEAE, CRYPTONEMIALES

Calcified, rose red segments tightly connected by flexible joints characterize this alga, which can reach a height of 6 cm. Branching is **dichotomous with narrow angles (paired branches almost parallel)**; consequently, all branches have an erect posture. Tightly packed clumps sometimes form small cushions on rock or dead coral substrates in quiet, calm, shallow waters, but specimens have been dredged from 30 m deep.

Haliptilon subulatum   (Ellis and Solander) Johansen
CORALLINACEAE, CRYPTONEMIALES

Small, compressed plants (1–3 cm tall), composed of brittle, heavily calcified, chalky pink segments connected by flexible joints. Segments of the **main axis and the larger laterals have three conspicuous ribs,** whereas tip segments are often cylindrical. Lateral pinnate branches arise from segments of the main axis. This species grows epiphytically on larger algae in tide pools or in the subtidal zone. Although the genus is found mostly in temperate waters, several small, inconspicuous species occur in the tropics. [*Corallina subulata* in Taylor (1960); see Johansen (1970) for update]

Amphiroa fragilissima   (Linnaeus) Lamouroux
CORALLINACEAE, CORALLINALES

Forming dense clumps or mats up to 8 cm thick of entangled, fragile, thin, brittle, jointed, calcareous branches, this species is generally yellowish to whitish pink. The dichotomous branches form unusually wide angles (broad Y's) at each joint with the **cylindrical branches swollen and enlarged at the segment ends.** Mats are lightly attached to the substrate in shetered locations, most commonly among seagrasses and in rock crevices 1–10 m deep.

Amphiroa rigida var. *antillana*   Boergesen
CORALLINACEAE, CORALLINALES

A relatively open, brittle, strongly calcified species with thin, narrow, **cylindrical branches (not swollen at the joints)** forming light off-white clumps about 10–15 cm in diameter. Although the species is dichotomously branched, the flexible joints seldom occur at the point of branching as they do in *Amphiroa fragilissima*. This alga is most commonly found in shallow waters (above 1 m), growing loosely attached among seagrasses.

Amphiroa brasiliana   Decaisne
CORALLINACEAE, CORALLINALES

Composed of pink, jointed, dichotomous, somewhat flattened branches growing close together, this plant forms low-growing (seldom more than 5 cm high) brittle clumps that are hard and stony because of their calcification. Generally **found in the crown or top of other sturdy plants** (often among the branches of the larger *Amphiroa hancockii*), but it can also occur on rocks 1–10 m deep.

Amphiroa tribulus   (Ellis and Solander) Lamouroux
CORALLINACEAE, CORALLINALES

Amphiroa tribulus has brittle, calcified, thin, flattened, sparse branches (2–4 mm wide) forming pinkish red, bushy clumps that may grow up to 10 cm tall. **Often, the edges of the branches are so flattened that the middle portion looks like a keel,** and the flexible joints occur most frequently where the branches divide in an irregular arrangement. Individuals are most abundant in crevices of the reef crest, spur-and-groove system, and patch reefs 1–10 m deep.

Amphiroa hancockii   W. Taylor
CORALLINACEAE, CORALLINALES

Plants are pinkish purple, with branches up to 1 cm wide that grow from an inconspicuous basal crust. **Branches are composed of flattened segments without a distinct midrib,** having irregular to somewhat dichotomous branching and commonly forming hard, stony (because of heavy carbonate deposits) heads or clumps up to 15 cm wide. It grows in rock crevices just beyond the reef crest in spur-and-groove areas (2–10 m deep) that receive considerable wave surge. Note that *Amphiroa brasiliana* also is present in the lower portion of this photograph.

Neogoniolithon spectabile   (Foslie) Setchell and Mason
CORALLINACEAE, CRYPTONEMIALES

Hard, stony plant forming knobby, calcareous, hemispherical (up to 15 cm in diameter) clumps that tightly adhere to rock or coral or, when broken free, accumulate with rubble in depressions. The knobs or pale pink protuberances vary in size and shape, and the branching is irregular to dichotomous. This alga is common in subtidal back-reef areas above 8 m in depth. [*Goniolithon spectabile* in Taylor (1960); see Adey (1970) for name change]

Neogoniolithon strictum   (Foslie) Setchell and Mason
CORALLINACEAE, CRYPTONEMIALES

Plants are heavily calcified, appearing hard and brittle, with irregular, tapered, cylindrical, blunt branching and no joints. The chalky, rose pink, thick branches (up to 5 mm in diameter) are rigid and stony, and they break or snap sharply when force is applied. Individuals form clumps 10–14 cm broad that consist of a mass of crisscrossing branches. A common and distinctive alga, **mostly upright and extensively branched, lying free on shallow seagrass beds** on the reef flat above 3 m in depth. [*Goniolithon strictum* in Taylor (1960); see Adey (1970) for an update]

Lithophyllum congestum   (Foslie) Foslie
CORALLINACEAE, CRYPTONEMIALES

Lithophyllum literally means "stone leaf," and *Lithophyllum congestum* forms rock-hard (from heavy calcification), **pink to purplish, branched, headlike plants** up to 15 cm high that are often confused with coral. Branches are crowded, stout projections (1–2 mm in diameter) emanating from the massive base and are occasionally broadened or waferlike. These plants are found in extremely heavy surf conditions on the reef crest, although broken fragments may occur deeper.

Mesophyllum mesomorphum   (Foslie) Adey
CORALLINACEAE, CRYPTONEMIALES

Distinctive alga consisting of dark red to pink **overlapping shelves or lobes** of brittle (especially on the edges) calcified crusts that spread flat against the substrate. Common in the subtidal zone, growing in shady cracks and crevices or epiphytically on other algae to a depth of 35 m. Some experts consider *Mesophyllum mesomorphum* to be a Pacific species and believe that this Caribbean plant, even though similar, is in fact a different species not yet assigned a specific epithet. [*Lithothamnium mesomorphum* in Taylor (1960); however, Adey (1970) moved several species of this genus to *Mesophyllum*]

Titanoderma sp. and *Fosliella farinosa* f. *callithamnioides*   (Foslie) Chamberlain
CORALLINACEAE, CORALLINALES

The thin pinkish gray crust on the green alga *Dictyosphaeria ocellata* is *Titanoderma* sp. It grows up to 5 mm in diameter as single patches or more often coalesces into crowded groups. The light-colored bumps (conceptacles) pictured here are the reproductive structures. The fine pinkish gray filaments growing over the surface of the green alga are *Fosliella farinosa* f. *callithamnioides*. Both of these small, inconspicuous, calcareous species grow epiphytically on many larger marine plants in the shallow (1–15 m) subtidal zone. [see Chamberlain (1983) for description]

Titanoderma prototypum   (Foslie) Woelkerling, Chamberlain, and Silva
CORALLINACEAE, CRYPTONEMIALES

Cream-colored to red crust with rather thin margins that develop in a circular pattern initially but later become more irregular in growth. The small hemispherical bumps that appear on the surface of the central portions of the crust are the reproductive structures (conceptacles). Plants are often found growing on dead shells or rubble fragments in shallow waters (less than 5 m deep). [*Lithophyllum prototypum* in Taylor (1960), see Woelkerling, Chamberlain and Silva (1985) for update]

Titanoderma bermudense   (Foslie and Howe) Woelkerling, Chamberlain, and Silva
CORALLINACEAE, CRYPTONEMIALES

This grayish to pale red, thick crust (up to 9 mm thick) consists of overlapping layers. The grayish lines on the surface are the result of the sloughing away of surface cells, which prevents faster-growing organisms from permanently attaching to the surface. This alga grows on calcareous substrates such as coral fragments in shallow waters, usually less than 5 m deep. [*Lithophyllum bermudense* in Taylor (1960), see Woelkerling, Chamberlain, and Silva (1985) for update]

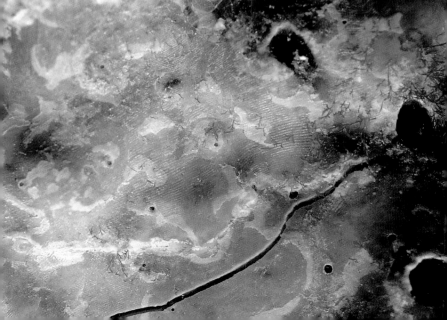

Peyssonnelia sp.
SQUAMARIACEAE, CRYPTONEMIALES

A commonly present crust, looking much like a **dark reddish maroon** coat of paint on rocks. Sometimes the extreme edges of the plants are raised above the substrate and not tightly adhering. Identification at the species level requires a microscope to reveal the infrequent, internal, reproductive structures. Common on hard substrates, this alga grows from the shaded intertidal region to extremely deep waters (200 m).

Porolithon pachydermum   (Weber-van Bosse and Foslie) Foslie
CORALLINACEAE, CORALLINALES

Calcareous (lime-secreting), **pinkish gray crust extensively covering intertidal reef-crest areas,** where it often contains regular holes, similar to Swiss cheese. The holes are caused by the chiton *Acanthochitona lata* Pillsbury, which feeds on competing upright plants that would otherwise overgrow and shade out the encrusting two-dimensional *Porolithon*. Also, this alga is perhaps the most important Caribbean reef builder. Because of its hard crustose form and tolerance of strong sunlight and desiccation, *Porolithon pachydermum* is one of the few organisms that can withstand the heavy surf and harsh environment of the reef crest, where it exists in abundance. It can also extend to a depth óf 10 m. Also pictured are several small pinkish purple patches of *Neogoniolithon* sp., another crustose coralline alga. [recent studies have noted no definable differences in certain Pacific species between the genera *Porolithon, Hydrolithon,* and the older genus *Spongites*—see Penrose and Woelkerling (1988) for details]

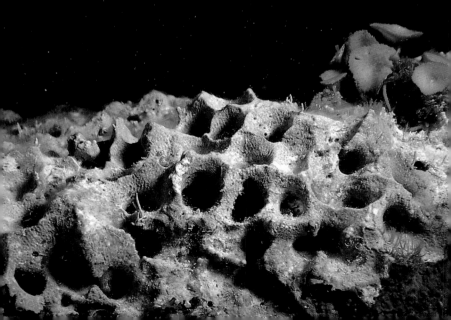

Sporolithon episporum   (Howe) Dawson
CORALLINACEAE, CORALLINALES

A hard, **calcareous, reddish brown crust** spreading in layers that often overgrow each other, these plants are generally characterized by irregular bumps or lumps with a smooth, glazed texture. The entire plant body can become 5 mm thick and cover extensive areas. When broken or chipped, the exposed surface is white, in stark contrast to the dark coloration of the undamaged plant. Crusts may be found on rocks or dead coral from the lowintertidal zone to 30 m deep. [*Archaeolithothamnium episporum* in Taylor (1960); see Dawson (1960) for genus change]

Hydrolithon boergesenii   (Foslie) Foslie
CORALLINACEAE, CRYPTONEMIALES

A calcareous crust that forms knobby rubble with a rough, chalky surface; easily recognized by its **purple-lavender color.** *Hydrolithon boergesenii* is most abundant on either side of the reef crest just below the low-tide line and on the seaward portion of shallow reef flats to a depth of 20 m. This species plays an important role in building reef structure by helping to cement coral fragments and loose rubble into a solid mass. [*Goniolithon boergesenii* in Taylor (1960); see Norris and Bucher (1982) for the update, and Penrose and Woelkerling (1988) for implications of the reestablished genus, *Spongites*]

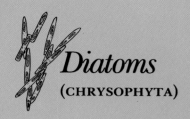

Diatoms

(CHRYSOPHYTA)

Diatoms are a group of silaceous (i.e., having glass shells), microscopic plant cells living either singly or in groups. When observable on the reef, they are usually in a bright yellow-green form consisting of separate cells embedded in gelatinous material to form chains. This gives the appearance of extremely fine filaments, and when many filaments combine, large (to 10 cm), soft clumps are formed. These clumps are generally found entangled on other plants, particularly *Acanthophora spicifera*.

Blue-Green Algae
(CYANOPHYTA)

The blue-green algae are not easily observed, consisting mainly of clusters of microscopic filaments that appear as dark (often black) patches, fuzzes, or crusts on rocky substrates. They are commonly found in the splash zone, where no other seaweed could exist because of extreme drying. Although cyanophyte coverage may be extensive, their low stature makes them inconspicuous, but they are slippery when touched (many intertidal rocks and boat ramps are hazardous to walkers because of the growth of microscopic blue-green algae). Marine blue-green algae are less prevalent than their counterparts in terrestrial and freshwater habitats. However, the role that these algae play in converting common atmospheric nitrogen into compounds that are usable as plant nutrients is extremely important on tropical reefs. Cyanophyta are related to the oldest known life forms on earth, and some consolidate sediments in a banded pattern to create permanent rock formations called stromatolites. Stromatolites date back to three billion years ago and are thought to have produced the critical levels of oxygen that led to higher forms of life.

Schizothrix calcicola   (C. Agardh) Gomont
OSCILLATORIACEAE, OSCILLATORIALES

This dark maroon, filamentous alga (each filament consists of a single row of cells) grows in matlike colonies that often trap bubbles of their own oxygen on bright sunny days. A truly ubiquitous plant found throughout the world's oceans to 30 m deep on nearly any substrate, this species is extremely important in converting atmospheric nitrogen into compounds that other marine plants can utilize as nutrients. Generally inconspicuous but often forming large gooey mats. [see Drouet (1963) for details]

Phormidium corallyticum   Ruetzler and Santavy
OSCILLATORIACEAE, OSCILLATORIALES

Black-band disease is caused by the alga *Phormidium corallyticum,* which is unusual in that it attacks and kills coral colonies. Filaments form a densely interwoven mat that appears on corals as a dark brown or blackish line, hence the common name. The coral is usually white (dead) behind the algal line, while its living color is apparent on the other side of the line. This is a fairly common disease in certain susceptible, shallow, reef-building species of coral. [see Ruetzler and Santavy (1983) for details]

Dinoflagellates
(PYRRHOPHYTA)

The dinoflagellates of the reef community are all microscopic and therefore not visible to the unaided eye. Zooxanthellae are microscopic, one-celled, photosynthetic dinoflagellates that grow profusely within the tissues of coral
animals, and they are partly responsible for the variety of
colors shown in the corals pictured here. The relationship
between the plant and animal components of the coral are
complex and not fully understood; nevertheless, it is sus-
pected that coral species possessing zooxanthellae cannot
survive without the alga for any extended period. Corals are
a large and diverse group of organisms; most reef-building
species contain zooxanthellae, and many non-reef-building
corals do not.

Flowering Plants

(MAGNOLIOPHYTAE, OR ANGIOSPERMAE)

Marine Angiospermae, unlike the algae, have true roots, stems, and leaves containing vascular tissues, as well as inconspicuous flowers that form spiny seeds. Even though relatively few marine species exist (about 50), they are often the conspicuously dominant organisms forming vast seagrass meadows. That contrasts markedly with terrestrial environments, where seed-bearing plants are the most successful in terms of both numbers of species and abundance. The predominant Caribbean angiosperm is the turtle grass *Thalassia testudinum,* often found growing with the manatee grass *Syringodium filiforme.* These seagrasses often form extensive beds on sandy and silty shallows. Such seagrass flats are important as breeding grounds for numerous small fishes and invertebrates.

Halophila decipiens   Ostenfeld
HYDROCHARITACEAE, BUTOMALES

Small, bright green seagrass (up to 5 cm tall) with a slender stem, generally composed of **two opposite, thin, oval leaves at each node** or segment of the slender rhizome. The leaf edge contains many minute teeth. Rhizomes run or creep throughout soft substrates, producing leafy patches up to 1 m across in moderately shallow waters (30 m deep) of quiet lagoons or brackish ponds.

Halophila englemannii   Ascherson in Neumayer
HYDROCHARITACEAE, BUTOMALES

At first glance, this seagrass resembles the alga *Caulerpa paspaloides* in overall aspect; however, the smooth, flat leaflets have distinctive midribs and veins. Slender horizontal rhizomes creep just beneath soft substrates and give rise to erect stalks that are **topped with a whorl of 6–8 small (up to 3 cm long), bright green, oval leaves** and may reach up to 20 cm tall. The edges of these leaves contain fine teeth. Normally found in shallow waters (less than 5 m deep), this seagrass grows on sandy or muddy bottoms of quiet lagoons, ponds, or boat harbors down to 40 m deep.

Syringodium filiforme   Kuetzing
POTAMOGETONACEAE, NAJADALES

Also a flowering plant, manatee grass is unique among sea-grasses in having **cylindrical leaves** rather than flattened blades. Its runners form dense mats in sandy to fine-mud sediments. Individual leaves are grass green and may reach 45 cm in length; they sometimes bear small, inconspicuous flowers and fruits. *Syringodium* is often found with *Thalassia testudinum* (turtle grass) in shallow-water meadows (1–10 m). *Syringodium filiforme* is the only species of this genus to occur in the Caribbean.

Halodule beaudettei   (den Hartog) den Hartog
CYMODOCEACEAE, NAJADALES

This grass green seagrass is similar in appearance to *Thalassia*, but the leaf is much narrower (2–3 mm wide) with slender flat blades (4–10 cm tall) connected by extensive mats of rhizomes or runners with relatively weak root systems. The most distinctive feature of *Halodule beaudettei* is the presence of **three teeth on the tip of the blade**; the middle tooth is generally much larger than the two side teeth. The flowers and seeds of this species have never been found, but in other species of this genus, the flowers are quite inconspicuous and fruits appear on the ends of long slender stalks. This seagrass can form meadows in shallow (1–10 m) waters over a broad range of salinities (salt levels).

Thalassia testudinum   Koenig
HYDROCHERITACEAE, BUTOMALES

Probably the most abundant marine plant in the Caribbean, turtle grass, as it is commonly known, has true flowers (greenish white to pale pink) and produces seeds (in a conspicuous pod). The development of the extensive root system allows expansion into new space by means of tough, well-anchored runners. The erect grass-green **leaves are 4–12 mm wide and can be as much as 1 m tall. Flattened strap-shaped leaves** have 9–15 parallel veins. The older blades are often heavily overgrown with epiphytic organisms. A ubiquitous plant of shallow (1–20 m) sandy or sediment bottoms, where it often forms lush meadows, *Thalassia testudinum* is the only species of this genus to occur in the Caribbean.

Underwater Photography and Collecting

Photography, for artistic and documentary purposes, is increasingly replacing the more traditional methods of collecting and preserving marine plants and animals. Marine plants are not only aesthetically pleasing in their natural surroundings—many are also essential parts of breeding and nursery grounds for marine invertebrates and fishes; all serve as food for various marine animals. Therefore, disturbing a community by removing and collecting large quantities of algae may affect more than just the plants. If you do collect, be selective in the specimens you take, and when photographing, be aware of corals and other delicate organisms that are easily damaged by trampling or careless diving.

Collecting and preserving algae can be time-consuming, but it is also rewarding when the purpose is to create scientific specimens as well as attractive wall mounts or decorations for stationery or greeting cards. That can be done, as it is for land plants, with a plant press (sheets of cardboard interspersed with absorbent blotters), heavy rag paper on which to press the alga, and wax paper to cover the specimen so it does not stick to the blotter above when under pressure from straps or weights. Always record pertinent information such as location, date, depth, substrate, and collector in the lower right corner of the paper. We recommend, however, that the casual observer record the wildlife of the reef with an underwater camera. Whatever type of camera is used, documenting marine life on film is a fascinating, challenging, and rewarding activity.

241

Camera choices are many, but the best underwater cameras today are amphibious units. The most popular of these (Nikonos) has features such as an automatic light metering system, extensive accessories, compact size, and automatic (through-the-lens) flash exposure. It is most attractive for a person who loves the out-of-doors (whether in or out of water) because it is rugged, durable, and dependable. The disadvantages are that the Nikonos is not a single-lens reflex (SLR) camera, so the photographer must compose through a viewfinder, not directly through the lens, and focus by measuring or estimating the distance. But even with those drawbacks, amphibious cameras are still the most dependable and compact units available for high-quality underwater photography.

Another option for quality pictures is a housed camera. Submersible plastic or metal housings are available through photographic and dive shops for most popular SLR land cameras, instamatics, polaroids, and video cameras. The housings are larger and bulkier than amphibious cameras and usually more fragile, but the results can be striking.

When just getting started in underwater photography, do not be too concerned about the type of camera, since a great picture is usually determined by the creative eye behind the lens and not by the equipment.

One of the most important elements in underwater photography is clear water. Water clarity is necessary for clear pictures; unfortunately, divers and snorkelers often do not have crystal-clear water available. Consequently, they have to work with what is typical, which all too frequently can be the murky greenish conditions caused by suspended matter. This particulate material (minute suspended particles of soil, fine sediments, decaying matter, or microscopic organisms) definitely affects the quality of the photograph in that it may reflect, scatter, or absorb light to varying degrees. Discolored water caused by dissolved substances may also result in undesirable effects. One thing to remember—if the subject does not appear clear, neither will the photograph. The closer the subject, the less water (and therefore less particulate matter) there is between the lens and the subject. Also, water acts as a filter and reduces contrast, so again the closer the subject, the better the contrast. The key reason many underwater photographers frequently use wide-angle lenses (15–35 millimeters) is to enable them to get close to a large area or object and still include the entire subject in the picture. In other words, reducing the amount of water between the camera and subject is always desirable.

Lighting is another critical part of underwater photography. Normal white light appears as such because it contains all the colors of the visible spectrum. Filters, commonly used in photography, absorb some colors and allow others to pass through. Ocean water functions as a large filter; the farther the light travels through water, the more pronounced the filter effect becomes (Fig. 9), until at great depths (400 meters in clear water) almost no visible light is present. Normal lens filters absorb a specific region, or color, of light, but water absorbs different colors at different depths and distances (selective absorption). In the top few meters of clear waters, cameras without electronic flash units (strobes) or other artificial light sources are effective, but as depth increases, the colors with longer wavelengths (particularly red light, Fig. 9), as well as overall light intensity, decrease. Therefore, in deeper waters, artificial light sources are essential to provide sufficient light and to restore accurate color and contrast to the subject. Artificial light can transform a mundane blue-gray scene into a vibrant array of color.

Using a strobe underwater, however, is not as simple as on land and presents its own problems, mainly stemming from the particulate matter in the water. When a subject is lighted, small particles in the water between the lens and the subject are also illuminated (Fig. 10). The result is a picture obscured by white specks where such particles have been brightly illuminated by the flash. Consequently, the seawater must be very clear, with reduced particulate matter, or something must be done to minimize the highlighting of the particles. We recommend positioning the electronic flash unit

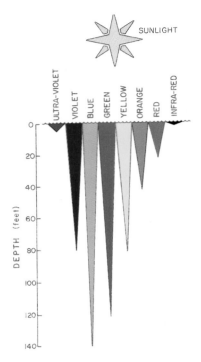

Figure 9. The differential penetration of the various light colors (spectrum) into clear oceanic waters.

Figure 10. The effects of placing the flash unit at increasing distances and angles from the camera. Note the reduction in particulate matter illuminated between the lens and the subject in the middle and lower drawings.

away from the body of the camera (instead of directly on the camera, as is commonly done on land), either by mounting the flash on a long adjustable arm or by holding it by hand. This concentrates the light on the subject rather than illuminating the water between the lens and the subject (Fig. 10). One other hint: since there is little color except blue in

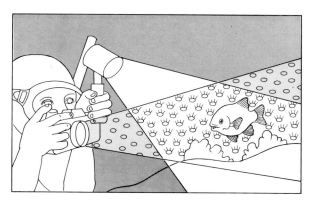

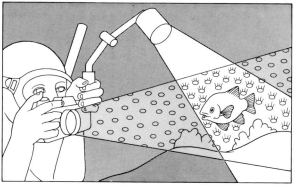

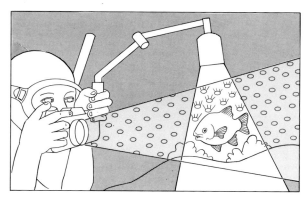

deep waters, it is often helpful to carry an underwater flashlight along when scuba diving to inspect various subjects before taking their pictures. The extra light allows the diver to assess the actual color instead of what the unaided eye sees—all blues and grays.

A nearly foolproof way to get dramatic, underwater, close-up shots of smaller plants is to use extension tubes (available at dive shops) or macro lenses and framers in conjunction with an amphibious camera. Extension tubes are merely watertight cylinders that, when placed between the camera and lens, enable one to focus on very close objects; the longer the tube, the closer the focal point (Fig. 11). When extension tubes are used with a single-lens reflex camera, focusing and framing are no problem, because the photographer can actually view what the film "sees" through the same lens. With a "blind" camera such as the Nikonos, the view is not through the lens but through a separate viewfinder, and the photographer cannot see the identical area as the lens in close-up photography. Therefore, extension tubes for most blind cameras come with metal framers that indicate the area to be photographed without need for a viewfinder. A framer consists of a metal bracket that connects the extension tube to a three- or four-sided frame set at a fixed focal distance from the lens. With the frame placed around the subject, the focal distance is always correct (camera lens set on infinity). Also, since the distance never changes with that specific tube and framer combination, once the lighting (with an electronic flash) is correct, it also remains constant.

Extension tubes have excellent resolution, since they do not affect the camera's optics, but they are not removable underwater, which can be a significant drawback. To avoid that inconvenience, front-mounted snap-on and screw-on macro lenses are available with the added advantage of being removable underwater. With those devices, the photographer can vary the subject matter on a single dive. The disadvantage is that when a lens (such as the snap-on macro)

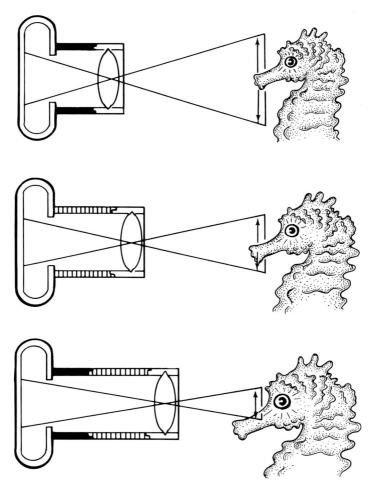

Figure 11. The effects on close-up photography obtained by adding various combinations of extension tubes (top—short tube, image ratio, 1:2; middle—long tube, 1:1; bottom—both tubes, 2:1, which doubles the size of the image).

is added to the camera's optics, some loss in the overall quality of the photograph is unavoidable. Each system has advantages and disadvantages; the optimum situation is to possess both systems.

Film selection always poses important questions in underwater photography, and in our opinion, the choice depends on personal preference. Higher-speed films (ASA/ISO 150, 200, 400, 1000) are useful for natural-light photography that requires quicker exposures (for instance, because of unavoidable motion underwater), but they lack the fine-grain quality of slower films (ASA/ISO 25, 64, 100). An artificial light source provides extra light, permitting the use of fast exposure speeds and slower films to gain sharper, fine-grained photos. Also, different types of film have various color qualities. For example, Ektachrome (Kodak) films usually have better blue rendition. For most natural-light underwater photography that is desirable, but quality in the red and green wavelengths is lost. Conversely, Kodachrome films have excellent red and green colors, so we prefer them when using a strobe, because the artificial light enhances those wavelengths to contrast with the ever-present blues of deeper waters.

If you are the type of person who appreciates a challenge, then capturing the artistic and natural beauty of the undersea world on film is extremely exciting. There is no terrestrial counterpart to underwater scenery. The vibrant colors and delicate structures of the plants and animals are unique to the marine environment. Sharing that beauty with friends and relatives after returning from an underwater adventure is rewarding, too. Although underwater photography is no snap, it can provide a lasting record of truly captivating experiences.

References

Abbott, I. A. 1984. Limu: An ethnobotanical study of some Hawaiian seaweeds. Third edition. Pacific Tropical Botanical Garden, Kauai, Hawaii.

Abbott, I. A., and E. Y. Dawson. 1978. How to know the seaweeds. Second edition. Wm. C. Brown Co. Dubuque, Iowa.

Abbott, I. A., and M. A. Doty. 1960. Studies in the Helminthocladiaceae. II. *Trichogloeopsis*. Am. J. Bot. 47:632–640.

Adey, W. H. 1970. A revision of the Foslie crustose coralline herbarium. K. Nor. Vidensk. Selsk. Skr. 1970(1):1–46.

Allender, B. M., and G. T. Kraft. 1983. The marine algae of Lord Howe Island (New South Wales): The Dictyotales and Cutleriales (Phaeophyta) Brunonia 6:73–130.

Baldock, R. N. 1976. The Griffithsieae group of the Ceramiaceae (Rhodophyta) and its Southern Australian representatives. Aust. J. Bot. 24:509–593.

Bliding, C. 1969. A critical survey of European taxa in Ulvales. II. *Ulva, Ulvaria, Monostroma, Kornmannia*. Bot. Not. 121:535–629.

Bold, H. C., and M. J. Wynne. 1978. Introduction to the algae: Structure and reproduction. Prentice-Hall, Inc., Englewood Cliffs, New Jersey.

Chamberlain, Y. M. 1983. Studies in the Corallinaceae with special reference to *Fosliella* and *Pneophyllum* in the British Isles. Bull. Br. Mus. (Nat. Hist.) Bot. 11:291–463.

Cheney, D. P., and G. R. Babbel. 1978. Biosystematic studies of the red algal genus *Eucheuma*. I. Electrophoretic variation among Florida populations. Mar. Biol. 47:251–264.

Cheney, D. P., and C. J. Dawes. 1980. On the need for revision of the taxonomy of *Eucheuma* (Rhodophyta) in Florida and the Caribbean Sea. J. Phycol. 16:622–625.

Dawes, C. J. 1981. Marine botany. John Wiley and Sons, New York.

Dawes, C.J., and H. J. Humm. 1969. A new variety of *Halimeda lacrimosa* Howe. Bull. Mar. Sci. 19:428–431.

Dawson, E. Y. 1960. New records of marine algae from Pacific Mexico and Central America. Pac. Nat. 1:31–52.

Dixon, P. S., and L. M. Irvine. 1970. Miscellaneous notes on algal taxonomy and nomenclature. III. Bot. Not. 123:474–487.

————. 1977. Seaweeds of the British Isles. Vol. I. Rhodophyta. Part 1. Introduction, Nemaliales, Gigartinales. British Museum (Natural History), London.

Drouet, F. 1963. Ecophenes of *Schizothrix calcicola* (Oscillatoriaceae). Proc. Acad. Nat. Sci. 115:261–281.

Earle, S. A. 1969. Phaeophyta of the eastern Gulf of Mexico. Phycologia 7:71–254.

Eubank, L. L. 1946. Hawaiian representatives of the genus *Caulerpa*. Univ. Calif. Pub. Bot. 18:409–431.

Fredericq, S., and J. N. Norris. 1985. Morphological studies on some tropical species of *Gracilaria* Grev. (Gracilariaceae, Rhodophyta): Taxonomic concepts based on reproductive morphology. Pages 137–155 in Taxonomy of economic seaweeds with reference to some Pacific and Caribbean species. Edited by I. A. Abbott and J. N. Norris. California Sea Grant College Program, La Jolla, California.

Gabrielson, P. G., and D. P. Cheney. 1987. Morphology and taxonomy of *Meristiella* gen. nov. (Solieriaceae, Rhodophyta). J. Phycol. 23:481–493.

Goreau, T. F., and E. A. Graham. 1967. A new species of *Halimeda* from Jamaica. Bull. Mar. Sci. 17:432–441.

Johansen, H. W. 1970. The diagnostic value of reproductive organs in some genera of articulated coralline red algae. Br. Phycol. J. 5:79–86.

Joly, A. B., and E. C. Oliveira. 1968. Notes on Brazilian algae. II. A new *Anadyomene* of the deep water flora. Phykos 7:27–31.

Lobban, C. S., and M. J. Wynne, editors. 1981. The biology of seaweeds. University of California Press, Berkeley, California.

Magruder, W. H. 1984. Reproduction and life history of the red alga *Galaxaura oblongata* (Nemaliales, Galaxauraceae). J. Phycol. 20:402–409.

McLachlan, J. 1979. *Gracilaria tikvahiae* sp. nov. (Rhodophyta, Gigartinales, Gracilariaceae), from the northwestern Atlantic. Phycologia 18:19–23.

Norris, J. N., and K. E. Bucher. 1982. Marine algae and seagrasses from Carrie Bow Cay, Belize. Smithson. Contrib. Mar. Sci. 12:167–223.

Oliveira, E. C., and R. P. Furtado. 1978. *Dictyopteris jolyana* sp. nova (Phaeophyta) from Brazil. Nova Hedwigia 29:759–763.

Olsen, J. L., and J. A. West. 1988. *Ventricaria* (Siphonocladales-Cladophorales complex, Chlorophyta), a new genus for *Valonia ventricosa*. Phycologia 27:103–108.

Olsen-Stojkovich, J. 1985. A phylogenetic look at selected genera in the Siphonocladales/Cladophorales complex using immunological data. Pages 59–63 in Proceedings of the Fifth International Coral Reef Congress, Tahiti, Vol. 5. Edited by M. Harmelin Vivien and B. Salvat. Antenne Museum-EPHE, Moorea, French Polynesia.

Page, J. Z. 1970. Existence of the *Derbesia* phase in the life history of *Halicystis osterhoutii* Blinks and Blinks. J. Phycol. 6:375–380.

Papenfuss, G. F. 1968. Notes on South African marine algae. V. J. S. Afr. Bot. 34:267–287.

Penrose, D., and W. J. Woelkerling. 1988. A taxonomic reassessment of *Hydrolithon* Foslie, *Porolithon* Foslie and *Pseudolithophyllum* Lemoine emend. Adey (Corallinaceae, Rhodophyta) and their relationships to *Spongites* Kuetzing. Phycologia 27:159–176.

Ruetzler, K., and D. L. Santavy. 1983. The black band disease of Atlantic reef corals. I. Description of the cyanophyte pathogen. PSZNI: Marine Ecology 4:301–319.

Schnetter, R. 1978. Marine Algen der karibischen Kuesten von Kolumbien. II. Chlorophyceae. Bibl. Phycol. 42:1–199.

Silva, P. C., E. G. Menez, and R. L. Moe. 1987. Catalog of the benthic marine algae of the Philippines. Smithson. Contrib. Mar. Sci. 27:1–179.

Sterrer, W. 1986. Marine fauna and flora of Bermuda. John Wiley & Sons, New York.

Sze, P. 1986. A biology of the algae. Wm. C. Brown Publishers, Dubuque, Iowa.

Taylor, W. R. 1960. Marine algae of the eastern tropical and subtropical coasts of the Americas. University of Michigan Press, Ann Arbor.

———. 1962. Two undescribed species of *Halimeda*. Bull. Torrey Bot. Club 89:172–177.

———. 1974. Notes on algae from the tropical Atlantic Ocean. VII. Rev. Algol. Nouv. Ser. 11:58–71.

Voss, E. G., et al., editors. 1983. International code of botanical nomenclature adopted by the Thirteenth International Botanical Congress, Sydney, August 1981. Regnum Vegetabile 111:1–472.

Voss, G. L. 1976. Seashore life of Florida and the Caribbean. Banyan Books, Miami, Florida.

Woelkerling, W. J. 1976. South Florida benthic marine algae. Sedimenta V. Comparative Sedimentology Laboratory, Division of Marine Geology and Geophysics, Rosenstiel School of Marine and Atmospheric Science, University of Miami, Miami, Florida.

Woelkerling, W. J., Y. M. Chamberlain, and P. C. Silva. 1985. A taxonomic and nomenclatural reassessment of *Tenarea, Titanoderma* and *Dermatolithon* (Corallinaceae, Rhodophyta) based on studies of type and other critical specimens. Phycologia 24:317–337.

Womersley, H. B. S. 1967. A critical survey of the marine algae of southern Australia. II. Phaeophyta. Aust. J. Bot. 15:189–270.

Wynne, M. J. 1986. A checklist of benthic marine algae of the tropical and subtropical western Atlantic. Canadian J. Bot. 64:2239–2281.

Glossary

ALGA (plural, algae)—a plant that is photosynthetic and reproduces by spores; algae lack true vascular tissues, flowers, and seeds.

ALTERNATELY BRANCHED—a pattern of branching in which one branch appears on one side of the stalk and, farther up, the next branch appears on the opposite side, alternating back and forth up the stalk (Fig. 3).

ARTICULATED—jointed; having a series of calcified segments separated by flexible uncalcified joints.

ASSIMILATOR—an erect photosynthetic portion that produces energy sources (carbohydrates) for algae.

ATOLL—a ring-shaped reef, surrounded by open ocean, enclosing a lagoon with no central high island, although low-lying carbonate islands may occur on the reef ring.

AUTHOR—person who named a given species, whose name appears after the Latin generic and specific epithets (the scientific name) of a plant; the last part of the complete scientific citation of an organism.

AXIS (plural, axes)—central line of an alga on which the parts are regularly arranged; the stemlike stalk.

BARRIER REEF—a reef that is separated from a land mass by a deep lagoon.

BASAL—toward the base or point of attachment.

BILATERAL—arranged on two sides in reference to a center line.

BIRD ISLANDS—islands where birds nest in great abundance; their droppings, acting as fertilizer, make the surrounding waters rich in nutrients.

BLADDER—vesicle or small saclike float that enables some algae to float or remain erect.

BLADE—the leaflike structure of an alga (Fig. 2), also called frond.

BRACKISH WATER—part seawater and part freshwater; diluted seawater.

BRANCHLETS—tiny or small projections off a main or heavy branch (Fig. 2).

CALCIFIED—having lime deposits (calcium carbonate, a chalky substance) within or on the plant; a heavily calcified (calcareous) plant has a stony texture.

CERVICORN—having unequal forked branching, with one fork being smaller and not dividing further.

CLASS—a group of related organisms forming a category ranking above order and below phylum.

COLONY—an organism composed of connected individuals.

CONCENTRIC—having curved zones that parallel the margin of a blade.

CONCEPTACLE—a raised or pitlike reproductive structure.

CRUSTOSE—forming a hard surface layer; covering as a crust.

CYSTOCARP—a dark red, swollen reproductive structure on female Rhodophyta.

DICHOTOMOUSLY BRANCHED—having each division or fork divided into two equal portions (Fig. 3).

DOMINANT—most abundant and conspicuous; having the greatest influence.

ECOLOGY—the study of the distribution and abundance of organisms in relation to their environments.

ENVIRONMENT—the total physical, chemical, and biological surroundings.

EPIPHYTE—an organism that lives on but does not get its nourishment from a plant.

ETHNOBOTANY—the study of how various cultures use plants.

FAMILY—a group of closely related organisms ranking above genus and below order, designated by the ending "-aceae."

FIBROUS—made up of coarse, threadlike fibers.

FILAMENTS—a slender, threadlike row of plant cells.

FORM—a variation (of external appearance) within a species caused by environmental factors; for example, bright-light versus shaded growth forms.

FRINGING REEF—a reef running parallel to and attached to the shoreline of an island or land mass that is not separated by a deep lagoon.

FROND—a blade or leaflike structure of an alga (Fig. 2).

GENUS (plural, genera)—a group of closely related organisms, usually consisting of more than one species; first Latin word of a scientific name.

HABITAT—the physical environment in which an organism grows; the place where an organism is typically found.

HERBIVORY—the consumption of plant material by an animal; grazing.

HOLDFAST—a rootlike structure or a disk that attaches an alga to the substrate (Fig. 2).

INTERTIDAL ZONE—area of the shoreline between the highest and lowest tidal levels.

IRIDESCENT—reflecting an interplay of rainbowlike metallic colors; glowing, shining (see *Dictyota bartayresii*—true color brown but iridescent blue).

IRREGULARLY BRANCHED—branching in no consistent pattern (Fig. 3).

LAGOON—a relatively deep (5–30 m) protected area behind a reef front; an offshore area separated from the sea by a barrier reef; the center portion of an atoll.

LATERAL—at, from, or toward the side.

LEEWARD—on the side downwind from the prevailing wind condition.

LINEAR—as in "linear frond," with the edges parallel; narrow, much longer than broad.

MACROSCOPIC—large enough to be seen by the unaided eye.

MANGROVE ISLAND—an island with dense thickets of mangrove trees around its perimeter; mangrove trees can live in seawater and help form islands in the tropics and subtropics throughout the world.

MICROSCOPIC—unable to be reliably observed without the aid of a microscope.

MIDRIB—the center veinlike structure of an algal blade (Fig. 2).

NOMENCLATURE—the system of naming organisms and the rules used to determine the valid scientific name of an organism.

NUTRIENT—an enrichment or fertilizer chemical essential for growth.

OPPORTUNISTIC—capable of rapidly occupying newly available space.

OPPOSITE BRANCHING—branching in two opposing directions at the same level on the stalk (Fig. 3).

ORDER—a category of taxonomic classification ranking above family and below class, designated by the ending "-ales."

PATCH REEF—an isolated reef pinnacle or mound in lagoons of barrier reefs and atolls.

PEAT—in the marine environment, the compacted remains of mangrove plants or other organic debris.

PHOTOSYNTHESIS—the mechanism by which plants, with the aid of chlorophyll, convert sunlight energy, water, and carbon dioxide into carbohydrates or food.

PHYLUM (plural, phyla)—a major division representing a separate evolutionary line within the plant kingdom.

PHYSIOLOGY—the study of vital processes (metabolism) and functions of organisms.

PINNATE—having branchlets close together on opposite sides of the main axis, in a featherlike arrangement.

PLANE—a flat surface, i.e., growth in one plane results in a two-dimensional plant.

PREDATION—the preying upon or eating of one organism by another.

PROLIFERATION—the forming of numerous new algal fronds on an older blade portion.

RECURVED—bent back away from the main curve of the axis.

REEF CREST—that part of a reef where the highest wave shock is released, usually an intertidal or very shallow subtidal area.

REEF FLAT—a shallow protected area behind (shoreward) the reef crest.

RHIZOID—a rootlike structure of an alga (Fig. 2).

RHIZOME—a horizontal stem, similar to the runners or stolons produced by strawberry plants (Fig. 2).

SACLIKE—appearing like a sack or bag; balloon-shaped.

SALINITY—salt content of water.

SEAWEED—a larger marine plant easily observed with the unaided eye.

SEDIMENTS—loose solid particles of matter that settle or sink to the bottom; includes sands, gravels, muds.

SERRULATE—like the teeth on a saw, with pointed projections on blade edges.

SPECIFIC EPITHET—the second Latin word in the scientific name given to a group (i.e., species) of very closely related plants or animals that can interbreed to produce fertile offspring.

SPHERICAL—globelike or ball-shaped; round.

SPUR-AND-GROOVE—an area seaward of the reef crest containing deep sandy grooves separating shallower, parallel, calcareous ridges (spurs) forming a comb-tooth pattern.

STIPE—the stemlike portion of an alga.

STOLON—a runner or rhizome (see rhizome, Fig. 2) connecting small upright fronds; horizontal stem.

STROMATOLITE—a geological formation created by blue-green algae that trap and consolidate carbonate sediments in a banded, domed pattern.

SUBSTRATE—the substance or surface on which an organism is growing.

SUBTIDAL—below the lowest low-tide level.

SUCCESSIONAL—pertaining to the sequence of organisms leading from newly available substrates.

SYMBIOSIS—the intimate association of two dissimilar organisms in a mutually beneficial relationship that is necessary for the long-term survival of each.

TENDRIL—a hooked or curled branch that entangles with other organisms.

TUBULAR—hollow and cylindrical; pipelike.

TUFT—a cluster of filaments, branches, or branchlets attached at a single basal point.

TURF—dense-growing, short, thick intertwined mat of small plants.

UNILATERAL BRANCHING—branches arise on only one side of the main stalk (Fig. 3).

UPRIGHT—the blade or branch of an alga that stands vertical to the substrate.

VARIETY—Latin name added to the generic and specific epithets of a species to designate a difference consistent within a species, but not significant enough to separate as a new species.

WHORLED BRANCHING—many branches (three or more) arising from one level on a main stalk (Fig. 3).

Taxonomic Index

259